MANAGING CHANGE

Creating competitive edge

DAVID TRANFIELD
and
STUART SMITH

IFS Publications, UK
1990

British Library Cataloguing in Publication Data
Tranfield, David
   Managing change.
   1. Manufacturing industries. Technological development.
   Implementation. Management aspects.
   I. Title   II. Smith, Stuart
   658.514
   ISBN 1-85423-085-9

© **1990 IFS LTD,** Wolseley Business Park, Kempston
Bedford MK42 7PW, UK

Produced for IFS Publications by Technical Communications.

Printed and Bound in Great Britain by
Short Run Press Ltd., Exeter, UK.

Acknowledgements

In preparing this book the authors would like to acknowledge the help they have received from many different quarters.

Much of the work reported here stemmed from two research grants undertaken by the authors and funded by the ACME (Application of Computers to Manufacturing Engineering) Directorate of the Science and Engineering Research Council (SERC), and by the Joint Committee of the Economic and Social Research Council and the Science and Engineering Research Council. Without this basic research, this book would not have been possible.

In managing these grants our grant co-holders, David Hughes at Polytechnic South West and John Bessant at Brighton Polytechnic, have had considerable influence on our thinking, and particularly John must share the credit (and the blame!) for some of the ideas in Chapter Four.

Our research fellows who have contributed extensively were Rosalie Kirkwood, Clive Ley and Paul Levy. Rosalie deserves a special word of thanks for her work on earlier drafts of the Cummins and Westland cases, and both Clive and Paul made significant contributions to the development of Chapter Four.

The three companies who have kindly agreed to us (and encouraged us in) publishing our work in their organisations also deserve our thanks. The optimism, enthusiasm and positive thinking which we saw at Cummins Engines, Westland Helicopters and Rotabroach made us realise what is possible with good management, and we are grateful that each of them wished to lend their name to the work done. Particularly we would like to thank Barry Mills at Westland, Ralph Stych at Rotabroach, and the late Ken Davies at Cummins for their ideas and support throughout our work.

Throughout its life, our Change Management Research Unit has had two secretaries, Glynis Cole and Patsy Wilson-Smith. Both of them have spent endless hours on drafts, as well as coping with the traumas of a change of operating system on our word processor. Our thanks to them for their patience and professionalism.

We would also like to thank the dozens of managers who have contributed to our ideas and thinking not only in companies reported here but in numerous other manufacturing concerns. Manufacturing lives in "interesting times", and it is with no little respect and admiration for manufacturing managers attempting to change their organisations that this book is written.

Lastly we would like to thank MCB University Press for their agreement in allowing us to publish material in Chapters Two and Three.

David Tranfield and Stuart Smith
Change Management Research Unit
Sheffield Business School

Table of Contents

CHAPTER 3
An Outline for a Methodology 33

CHAPTER 4
Manufacturing Organisation for the 1990s 49

CHAPTER 6

Cummins Engine Company 113

CHAPTER 1

Background and Theoretical Framework

This book focuses on the managerial and organisational problems of implementing technological change in manufacturing companies. Its major thesis is that successful implementation requires a re-examination in management thinking in many companies. New manufacturing strategies are needed in the light of the opportunities for quantum leaps in competitiveness offered by new manufacturing technology and production management methods. These are often best handled by implementing changes on a wide front throughout the company in order to facilitate optimal exploitation. This chapter outlines the basic problem facing many manufacturing managers, specifies the main basis on which our conclusions are drawn, and outlines the theoretical perspectives which can be usefully brought to bear.

The Problem

The introduction of new technology is widely seen as one of the major means of enhancing the competitiveness of British manufacturing in world markets. The application of computers to design, production, assembly and management in manufacturing facilities offers opportunities, for example, for increasing flexibility and quality, and for reducing costs, work in progress, inventories and lead times. Unfortunately many of the applications of computers in British industry are not successful. Reviewing the literature reveals two major problems reported in implementing AMT:

(a) Companies experienced that many of the benefits were gained prior to installation. This varied from study to study (Dempsey (1983) 40%, Bessant & Haywood (1985) 75%, McCracken (1986) 90%), and

(b) Approximately 50% of applications were seen as unsuccessful (Ingersoll Engineers (1985) 50%, Voss (1985) 57%).

Both of these struck us as surprising in a sector with years of experience of implementing change. What is so difficult about implementing new technology?

There is widespread agreement that this low success rate is the result of difficulties in implementation, rather than inherent problems in the technology itself. What is more, the most significant implementation problems are seen as being associated with the management and organisational issues involved rather than technical commissioning or resistance at a shop floor level.

We are attracted to this view after having held many discussions with a variety of manufacturing companies, consultants and suppliers of new technology, which fleshed out the implementation problem. New technology by itself does little to increase competitiveness; it is how it is exploited that counts. Computers applied to existing facilities and systems merely make them more efficient. Substantial changes in competitiveness are only achieved by management fundamentally rethinking how new technology can be exploited in pursuing business strategy. Frequently, changes in approach are required at all levels of management.

Particularly at senior levels, managers need to recognise that the application of new technology can fundamentally change the competitive edge of the company. This involves thinking through the business strategy of the company and relating it clearly to manufacturing as a main contributor to competitiveness. In turn, senior manufacturing management needed to think through how changes in flexibility, quality, delivery, WIP, inventory, cost and reduced lead times might impact on business performance and affect other functions. Often this involves manufacturing management in reappraising its strategy on, for example, what to make and what to buy, which capital equipment to invest in, what kind of relationship to develop with suppliers, which parts to rationalise, the policy on design for manufacture, factory layouts, inventory policy, production management methodologies, etc.

All of this is leading to a much closer relationship than has existed in the past between manufacturing and other business functions. It means that manufacturing is becoming a key resource in business competitiveness, and we believe, it explains the two surprising research findings reported earlier, in that where this relationship was thought through, the company benefited prior to installation, and where it was not, the application was later seen as a failure.

However, our discussions with managers also suggested that it was not just a matter of policy that was critical in exploiting new technology. Senior management needed not only to understand and be committed to capital investment in new technology and its business effects, but also to think through and support the organisational and managerial changes necessary for its implementation. Frequently this meant bringing together functions which had for years existed separately, for the new technology was not respectful of traditional boundaries. As we talked to more and more companies about their experience, it became clear that the second industrial revolution is more than simply a technological revolution. It is demanding a revolution in how managers think and in how they organise manufacturing.

The argument that a shift in strategic thinking regarding the role of manufacturing is needed, is not novel. Many writers in recent years [e.g. Schonberger (1986), Skinner (1985), Hayes & Wheelwright (1984), Hill (1985) and Ettlie (1989)] have argued that the demise of Western manufacturing companies has reflected their inability to rethink the role of manufacturing as a strategic competitive tool. This is corroborated, in our experience, by the fact that despite these exhortations and the wide usage of the term 'manufacturing strategy' in the literature, very few companies explicate one in practice.

The pervasive feature of all applications of information technology is that they facilitate much greater and quicker integration and control of the manufacturing process, and it is these features which provide opportunities for quantum leaps in competitiveness. Indeed, we came to the conclusion that it is the increased speed and degree of integration of various activities within the business, created by computer applications, that is at the core of why new technology constitutes the very basis of the second industrial revolution. Unfortunately, in our view, this is both the promise and potential downfall of AMT applications. CAD/CAM systems, for example, can dramatically reduce lead times through increased speed of design and production, and also through the increased integration on a common database of the design, production engineering and production stages of manufacture. But, in order to achieve these benefits, it requires management to look at the performance of the whole design-to-production system. It seems only too frequent that CAD systems become merely more efficient electronic drawing boards, and CAM mere NC tape-controlled machine tools. There is little point in making the design process more efficient if it simply creates bottlenecks in either production engineering or in manufacturing departments. To achieve greater throughput and reduced lead times, it needs the whole of the design-to-production system to be optimised. Necessarily, this also involves a re-examination of the way these areas are managed and organised. The CAD/CAM information system cuts across traditional organisation boundaries and frequently problems in successful implementation are due to failure to recognise that new organisational realignments are needed.

Similarly, CAPM systems provide both promises and problems. They provide the capability for the more rapid integration of the planning and control of the whole manufacturing process with the consequence that it is possible to achieve, for example, significant improvements in inventory, WIP and throughputs. However, in order to deliver this, it usually requires the integration of a wide variety of information systems which have traditionally been the 'property' of different groupings in the company. Frequently, not only are there technical information system problems in integrating these systems, but there are also managerial and organisational problems in developing the 'collective will' to make these integrated systems work at the level of data capture and update.

Clearly, fuller integration between the manufacturing system and other business functions as well as with suppliers is both made possible by new technology and is a requirement for its successful competitive exploitation. Indeed, Waterlow and Monniot (1986) have found that the greater the level of integration demanded by new computer-aided production management technologies, the greater the difficulties of implementation, whilst Ettlie (1988) outlines his 'synchronous innovation' approach incorporating equivalent technological and administrative changes for effective change. All of these arguments point to the need for the development of a strategic methodology for guiding the thinking of management when implementing new technology, and to ascertain the main parameters which are to dictate manufacturing organisation to enable full exploitation. Both of these areas are the subject of this book.

A Theoretical Framework

In examining the area of managing change in manufacturing, we had to address the issue of what constituted 'success' in implementation. We decided to take a 'user-led' definition. Successful implementation, in our terms, was a multi-dimensional issue and was defined to us by managers as containing four major components:

    (a)   Business validity - had the new technology contributed to the business-related improvement (increased quality, decreased cost, decreased lead time, etc.) for which it was purchased, i.e. was it worth it? This was of vital concern to our collaborating managers, yet only rarely did we find a clear relationship between corporate strategy, manufacturing strategy and new technology or methods. It was as if the technology were assumed to deliver benefits irrespective of the context in which it was applied. We termed this *'the exploitation problem'*.

(b)   Technological validity - was the hard- and soft-ware installed, running and delivering according to specification, i.e. did it work? Frequently we found that although installing advanced manufacturing technology and particularly CIM systems poses a substantial technical challenge, it is not an insuperable problem nor is it seen by those involved as the most awkward problem. Frequently, technological installation was confused with the wider issues of implementation, involving not just project managing the introduction of new technology but also instigating and managing the adjacent business and organisational changes. To highlight the distinction between installation and implementation, we termed this issue, *'the installation problem'*.

(c)   Organisational validity - was the organisation designed or redesigned in both structural and cultural terms to deliver the benefits? We termed this *'the change problem'* to highlight the issue of companies developing appropriate organisational changes to match the requirements of the new technology. Rarely was this done in a consistent and systematic way, yet managers reported to us that this area contained some of the most difficult and intractable features of the implementation process.

(d)   User validity - did those who operated the new technology use it, have the skills to use it, and feel comfortable using it? Even if c) above is handled well at the level of organisation design the new technology has to be introduced to users in an appropriate way, and jobs redesigned as necessary. Hence we termed this *'the introduction problem'*. Sometimes this area was addressed by managers who usually employed 'human factors' specialists to help.

Overall, our work has pointed out the need for managers to distinguish strategies in each of these areas. Rarely did we see this happen, yet when it did success in introduction was almost guaranteed. The process was complicated requiring change on a wide variety of fronts all at once, but clearly this approach reflected the requirements of this technology. Later in this book, we suggest three dimensions which are vital for successful introduction of new technology (we amalgamate c) and d) above into an organisational dimension) and argue that an integrated change strategy is needed if companies are to innovate successfully.

Although some companies had encountered some technical problems in installation and commissioning, most, like those cited in other research, did not regard these as crucial in determining overall success. Frequently, companies reported difficulties with suppliers of equipment and the problems of operating at the forefront of technological development, but these seemed to be problems with which most of them were familiar, and had experience of handling.

By far the most significant issue that pervaded our discussions was the notion that the exploitation required a radical change in how management think and organise manufacturing. To do this seemed less difficult in greenfield site applications, in small companies, and in situations where there had been some powerful external threat such as takeover, merger or collapse of the company. The effect of each of these is to make radical change more palatable.

In our opinion, the key determinant of success was the way in which the companies had tackled the issue of managing change, and particularly whether they had managed to break the inertia of traditional taken-for-granted ways of thinking. If they had seen the implementation as a step function change and had adopted proven methods suitable for achieving this type of organisation-wide change, then they were likely to succeed. However, if they had simply seen their investment as a way of merely making the existing organisation more efficient, then they were unlikely to reap the potential rewards.

In addition there were those who recognised that new technology offered opportunities for quantum leaps in competitiveness, but did not recognise that this could only be achieved by quantum leaps in management thinking and organisation. These companies tended to attempt to initiate and manage changes using ideas in line with their previous experience of implementing technology. That is, they regarded new technology as creating small-scale knock-on effects which were best coped with in piecemeal fashion, as they emerged. However, success was seen as 'patchy' with this approach, and again we were reminded of the findings reported in (a) on page 2. It appeared that perhaps it was the radical rethinking of manufacturing, often stimulated by the technology, that was producing many of the benefits, rather than just the application itself.

The main point seemed to be that the impact of applying micro-electronics and revised production methodologies to manufacturing companies was so extensive as to require changes toward a different order. It was not simply making manufacturing more efficient, but was transforming what was possible, and in doing so challenging not only existing custom and practice, but also placing new demands on the capacity of management to manage step function change.

Theory of Change

Because the implementation of new technology involved organisational transition, the main theoretical perspective that we brought to bear was taken from the theory of change. This is a relatively new and diffuse field, but Smith (1982) outlines two distinct ways in which

change may be concept-
ualised [see Fig.1.1].
Firstly, there is morpho-
static change which
preserves an order by
treating disturbance as
external noise requiring
minor adjustment or
blocking out. Change in
this morphostatic sense is
therefore incremental.
Secondly, there is
morphogenic change which
treats disturbance as
information about internal

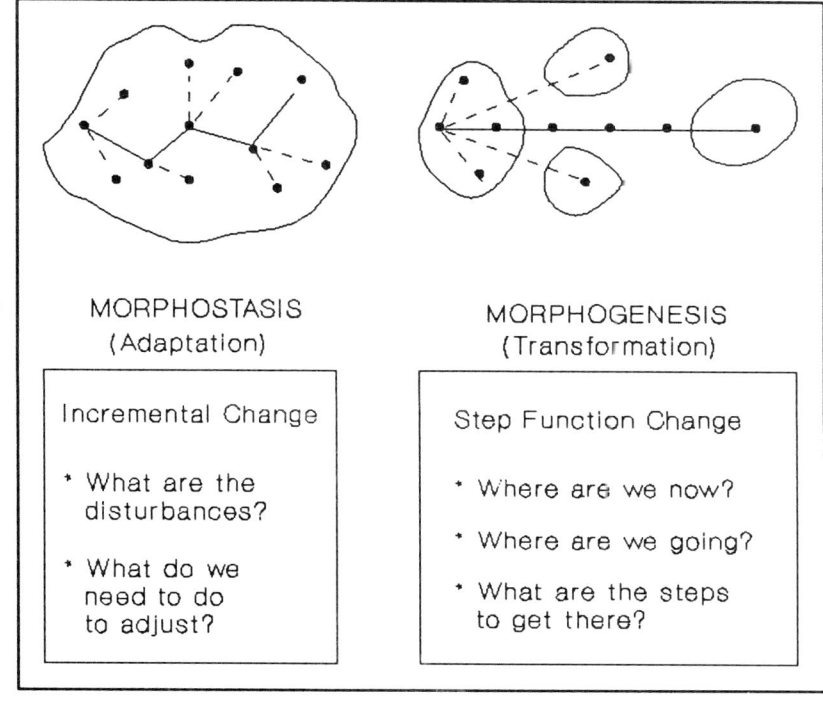

MORPHOSTASIS
(Adaptation)

Incremental Change

* What are the
disturbances?

* What do we
need to do
to adjust?

MORPHOGENESIS
(Transformation)

Step Function Change

* Where are we now?

* Where are we going?

* What are the steps
to get there?

Figure 1.1 Models of Change

conditions and suggests that the system should respond by altering orders. In this way,
change in a morphogenic sense produces a logically different order than that which came
before.

Our belief is that this simple framework possesses much credence in explaining the difficulties
being encountered in implementation. In the past, technological change in organisations has
been introduced by managers usually on an implicit or explicit morphostatic model. This has
been effective where the technology requires only an incremental readjustment, such as a
limited redesign of jobs or some skills training, within an overall system which remains the
same. However, where the technology does not just make the existing system more efficient,
but transforms what is possible and has implications for the organisation on a wide variety
of fronts including corporate strategy, organisation structure, job design, management
attitudes, etc., then clearly a morphostatic model becomes inappropriate. What is required
is that management do not just go through the morphostatic loop:

- what are the disturbances?
- what do we need to adjust?

but instigate a change programme inside the system designed to question the basic order and bring about a revolutionary or morphogenic change. Typical planning questions in this situation become:

- what is our current situation?
- where are we going?
- what are the steps to get there?

This suggests a 'vision-building' approach to change requiring strong leadership and a clear view of the horizon toward which the company is moving. Morphogenic change can only be planned and achieved when such a 'vision' of the future has been developed. The actual change itself becomes a series of incremental steps toward a 'vision of the future', thus 'smoothing' the step function.

Morphogenic change is almost always seen as uncomfortable by those inside the system. This kind of change generates high levels of insecurity. In this sense, one can understand managerial reluctance to inaugurate morphogenic change. Step function changes, by their nature, signal the overthrow of existing organisational arrangements, and involve the transition to a new order. Individuals may end up with new 'empires', reduced 'empires' or no 'empires' at all, which can lead to caution among managers, and the need for new skills and work arrangements can lead to resistance elsewhere in the organisation. Not only this, but the embeddedness of most managers within the existing system often makes it difficult for them to develop sufficient distance to diagnose the extent and scope of the changes required and how these might be achieved. It is not surprising that fundamental change is so difficult to achieve and that organisations contain so much inertia for continuing as they are.

Unfortunately, as many appear to have found out to their cost, computers by themselves do not create morphogenic change - the second industrial revolution. Computers applied to existing facilities and systems merely make them more efficient, achieving morphostatic change. Quantum leaps are achieved only by management fundamentally reappraising how new technology can be exploited in enhancing the competitive position of the company, and defining the technological requirement to facilitate this, whilst at the same time redesigning the organisation where required. Finally, the technology, if required at all, can be installed and commissioned. It is our conclusion that most companies achieving successful implementation on the four dimensions outlined previously could be seen as implicitly or explicitly using a morphogenic change model rather than a morphostatic model. Furthermore, these companies were prepared to view the implementation issue as multi-disciplinary, and to distinguish between implementation (including user and business validity) and technical installation. Lastly, these successful implementations rethought business operations first and considered technology last in the implementation process, alongside an organisational redesign as shown in Fig.1.2.

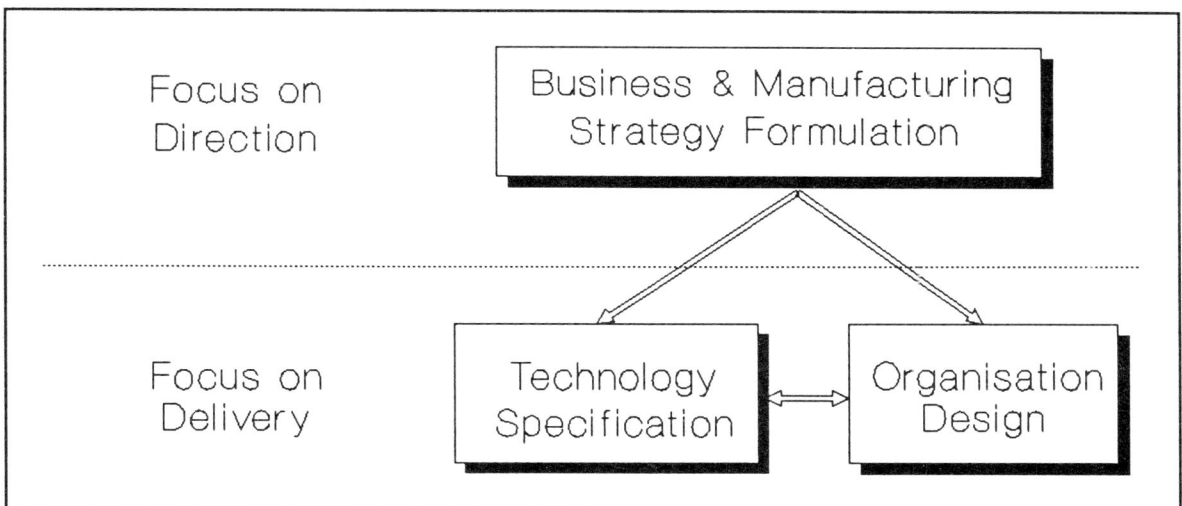

Figure 1.2 Focus for Implementation

We found this approach to thinking about different models of change useful in making sense of what companies were telling us about implementation, and it also provided the theoretical underpinning of an outline methodology, the aim of which was to bring about planned morphogenic change - 'a smooth transformation'.

CHAPTER 2

The Implementation Cube[1]

This chapter argues for a strategic view of implementation as a key process in the effective development of advanced manufacturing systems. Change has to be pursued not only in terms of technology, but also in terms of the associated organisational and business dimensions. Failure to do this can severely limit the impact and success of the application to the business in question.

Throughout the 1980s there has been a growing realisation amongst practitioners, consultants and researchers of the multi-faceted nature of advanced manufacturing systems. At no point in the development of advanced manufacturing systems has this been more keenly felt than in implementation. It seems that in the analysis and design phases of development, it is often possible to proceed on the basis, indeed the fond belief, that the major hurdles to be surmounted are technological and can be addressed solely using engineering or computing knowledge and skills. It is not the intention of this chapter to devalue or downgrade the

---

[1]This chapter is based on 'The implementation cube for advanced manufacturing systems' - *International Journal of Operations and Production Management*, 9,8, 1989, MCB University Press.

considerable technological challenge which advanced manufacturing systems pose. It is the intention of this chapter to argue that a purely narrow technological perspective is insufficient, producing sub-optimal implementations, and that a wider focus is needed if such systems are to be implemented effectively. This wider focus needs to encompass the business strategy and organisation design dimensions. In this sense, integration is required with both the business and organisational context in which the technology is to fit, as well as within the computer applications themselves.

The three dimensions of business, technology and organisation, constitute conceptually different aspects of the technological innovation process, and in implementing advanced manufacturing systems, it is useful to represent these dimensions as three orthogonal dimensions from which at least eight logical positions can be explored. Although several of these have strengths, all have logical and practical weaknesses, as the illustrative material from our cases shows. An argument has been presented therefore for the development of a strategy containing all three dimensions considered in the order: business first, technology and organisation afterwards, which aims to bring about transformational change on a wide variety of fronts to support the effective implementation and exploitation of the technology involved.

Dimensions of Implementation

From our observations we believe that in successfully implementing advanced manufacturing systems and technologies, managers must develop an overall change strategy based on the concern they have for three important dimensions: the business dimension, the technological dimension, and the organisational dimension. We see these dimensions as best being represented as orthogonal dimensions of a cube. Their orthogonality reflects the distinctive perspective each brings to the problem of implementation, and the 'cube' their inter-connectedness and inter-dependency.

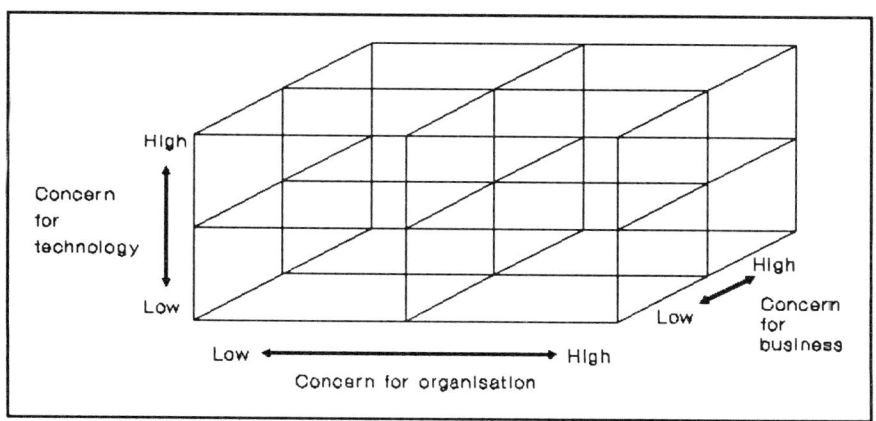

The Implementation Cube

It is our conclusion, that successful implementation involves devising an overall strategy explicitly attending to all three of these dimensions. Whilst each dimension is important in its own right, the linkages and inter-dependencies between dimensions are vital if an integrated technology is to be implemented effectively. The cube identifies positions in which managements may find themselves as they show more or less concern for each of the three dimensions. It is important to note that in our view it is the perceptions of management of each of these dimensions which dictates the level of concern. In this sense, we follow the meaning of 'concern' offered by Blake and Mouton (1972) in discussing concern for the dimensions of management. They make the point that:

'"concern for" is not meant to indicate how much (of the dimension under discussion) .... rather the emphasis here is on the degree of "concern for" which is present because actions are rooted in (it) and flow out of (these) basic attitudes' (p.8)

In this particular case, the actions flowing out of managerial "concern for" will be reflected in the planning arrangements made for innovation in each of the three dimensions making up the cube. However, as we argued in Chapter 1, our work suggests that chronologically the business dimension should be addressed first and the other two dimensions thereafter.

15

However, this ordering should not be interpreted by the reader as suggesting a reflection of relative importance. For technological innovation to be successfully implemented it is our view that management should place equal weight (i.e. have equal 'concern for') each of the three dimensions outlined, and develop an overall strategy with separate, but related strategies for each dimension.

It may be useful at this point to spend some time outlining further the meaning and scope of each of the dimensions:

Concern for Business Dimension

This refers to the overall strategic direction and competitive position of the company. It involves senior management in scanning the environment of the company, particularly with reference to customer expectations, potential and future markets (post implementation) and competitor activity. Of particular interest is the need to ascertain the precise nature of competitive edge in the market place (e.g. lead time, cost, quality, delivery, etc.) for this will determine the need to invest in new technology, and the size and scope of the investment, and set the success criteria by which the investment is to be judged. The development of this strategic business view into a manufacturing strategy designed to deliver it is equally important.

Concern for Technology Dimension

This refers to the development of an installation strategy for the investment. In practice, although most manufacturing companies find this a demanding problem it is probably the best resourced, most acknowledged and most clearly thought-through dimension. The development of manufacturing systems can be divided from a technical viewpoint into the development of a strategy for the installation of manufacturing technology and the development of a strategy for the development of the associated manufacturing methodologies and information systems. The former needs to identify the specific manufacturing systems architecture and establish clear benchmark tests against which to judge the performance of the hardware. The latter requires clear definition of systems interfacing between manufacturing and other business functions such as design, engineering and procurement/supplies, as well as the introduction of revised manufacturing methods such as JIT, TQM etc. The whole installation process usually requires clear project management of both of these aspects. Steps along the road to installation are well-documented and usually involve procedures which investigate alternatives/feasibility, develop and test the new system, install and commission the new system, and then hand it over to the user.

Concern for Organisation Dimension

In contrast to the technological dimension, this is probably the least well thought-through dimension of the whole process. Many companies with whom we have worked define organisational problems as 'noise' in the system and tend to relegate them to 'crisis management' or 'mopping up' activities after the main installation, rather than seeing them as an inherent and vitally important part of the whole socio-technical design and implementation process.

Broadly speaking the organisational dimension can be divided into two major areas, job design and organisational design. Traditionally, job design has tended to dominate the area and has come to be associated with the term 'human factors'. However, an increasingly important aspect of integrated technologies concerns the need to review and redesign manufacturing organisation at both the structural and managerial levels to match the demands of the technology, which usually has the effect of breaking down old departmental and functional allegiances, and creating new organisational groupings which are more compatible with both business goals and integrated technological systems.

For example, Parnaby 1988 has suggested three structural groupings relevant to the organisation of modern manufacturing companies: firstly marketing, product development and manufacturing systems design, secondly manufacturing operations and services, and thirdly finance and administration. If this framework were adopted, it would constitute serious reorganisation of most UK manufacturing companies and a major realignment of structural arrangements between functions. Chapter 4 contains a fuller exposition of the structural realignments necessary to adopt advanced manufacturing systems.

A second requirement is the need to change the culture of most manufacturing enterprises to parallel the needs of the technology and business. Organisational culture, as will be argued in Chapter 3, is seen as those shared, taken-for-granted assumptions and values which underpin the day-to-day operations of the company (Schein 1984). Most of the time, culture is not seen as a predominant issue for managers and engineers because in times of relative stability it is part of the background assumptions that no-one questions. The radical changes required in fundamentally rethinking manufacturing organisation in order to fully exploit advanced manufacturing systems questions some of the taken-for-granted assumptions which constitute the company's culture, and hence it becomes a focal issue for change. For example, traditionally, manufacturing organisations have been highly differentiated between design and manufacture. Integrated technologies, such as CAD/CAM or MRP II, immediately raise the issue of interdependency and identification with the corporate whole as a base requirement for success. In our experience this is a very difficult issue for most manufacturing companies to recognise and understand, and, having done so, to develop a strategy for change. Yet it is often of critical importance.

Exploring The Cube

In our work within manufacturing, we have come across many examples of different kinds of implementation strategies. In representing these within the framework of a cube we have reduced a large number of alternatives down to a manageable size, and in generalising necessarily cannot capture the exact nuance of local applications. We believe this to be appropriate given the nature of our task is to provide a strategic framework giving overall guides to thinking about and formulating technological change. It is our intention in the rest of this chapter to explore different segments of the implementation cube, exploring the strengths and weaknesses of the strategies located in the various segments. To facilitate this process each of the three dimensions is divided into a low and high category. This produces eight possible positions reflecting the emphasis placed on each of the dimensions both during the pre-planning and active phases of implementation. The descriptions of the eight segments of the cube are synthesised typologies based on a number of case studies.

1.   Low Concern for Technology, Business and Organisation

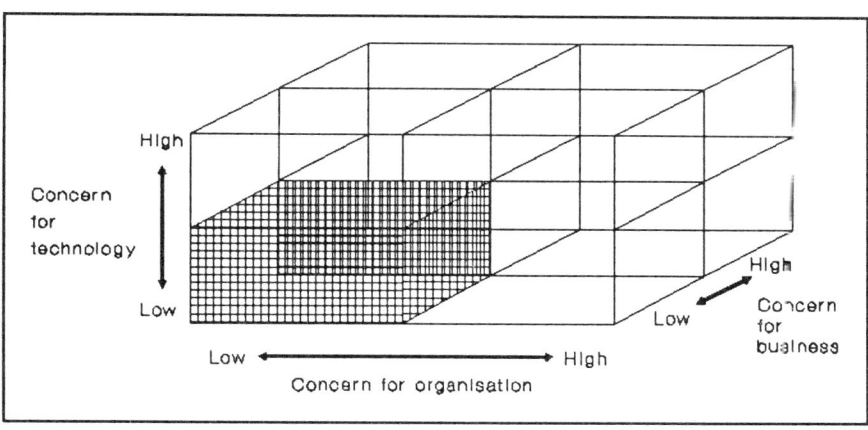

This constitutes the most poorly thought-out of all strategies available for implementation. There is little commitment and ownership from either the line management of the company as a group or from the specialist technologists. The idea for innovation is often a function of chance encounter (articles in journals, calls from sales people, contacts with other companies, etc.), sometimes coupled with an availability of finance prior to a period end. There is no real planning of how the new technology is to link into the business strategy or how the company organisation needs to be adapted to cope. Examples of this strategy are few and becoming fewer in integrated technologies, but sadly still can be found. One example which we found was in a heavy engineering company which had decided to invest in several areas including both design (CAD) and in manufacturing (CNC and FMS). In each of these areas, the choice of technology was providing only limited increases on current performance. For example in the CAD area, a serious shortage of plotters had severely limited use of the technology, and on the shop floor, the FMS suffered from lack of maintenance and services due to suppliers going out of business a week after its installation. No reorganisation had been planned to take account of these technologies with the result that only the extremely user-friendly CNC machine tools were accepted and used appropriately by the shop floor. The company had severe business problems at the time due to cutbacks from customers whose main business was operating offshore in the North Sea. Sadly, the investments in advanced manufacturing technology were not fitted to delivering a strategy for rejuvenation. Altogether, none of the above dimensions appeared to drive these disastrous investments, and exploitation and acceptance was left to chance.

Unless the company is particularly fortunate in finding a chance matching between business need, organisational acceptance and technological capability, the innovation will fail. At best the adoption of this strategy means that the application will be implemented late and/or deliver only part of its technical capability. At worst, it will be allowed to stand unused in a corner of the factory.

2.   High Concern for Technology/Low Concern for Business and Organisation

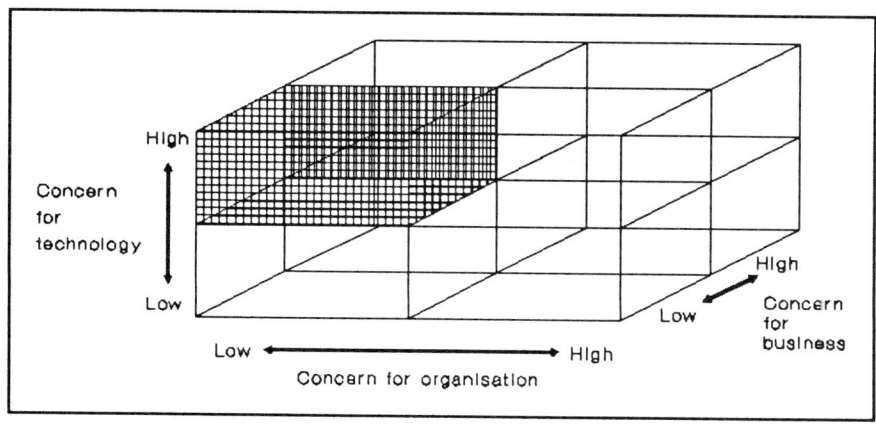

This is a typical technology-led strategy dictated by engineering or computing or a combination of the two. Usually driven by notions of technical excellence, there is an air of naive optimism about this strategy. It tends to be championed by technologists who require and authorise sometimes large capital expenditures on items marginally related to the development of competitive edge and for which there is no organisational implementation plan. It is as if the technology is assumed to deliver enhanced performance by itself, context-free. A good example of this was a heavy engineering company planning to invest several million pounds in a CAE system in the hope that this would revive competitive fortunes. The original specification included 3-D modelling facilities on all terminals, which preoccupied the steering group. There was only a minor organisational change programme outlined to facilitate the introduction of the technology, and this led to both official and unofficial action among the engineers when it arrived. Although seen as a technology to revitalise the company, when company products were costed, only 15% of the cost was attributable to design, and only some 3-4% to CAE'able items. Fortunately, this company after struggling to introduce the technology, re-assessed its position and was able to reassess its needs, respecify requirements and learn from its experience before too much damage was done.

The usual outcome of this common strategy is disappointment. This can take two forms. Firstly, the implementation will be extended because of the lack of organisational change programme involved, and secondly there is usually a late or more often limited pay-off in business terms.

3. High Concern for Technology and Organisation/Low Concern for Business

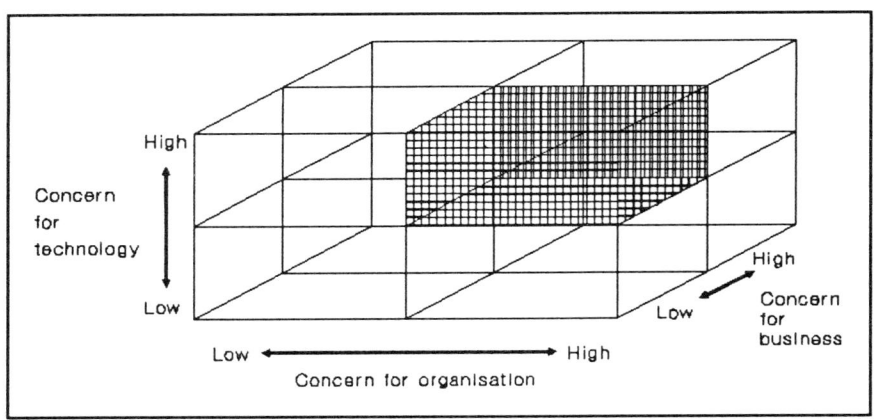

This constitutes the next step for many companies from 2 above. In learning that technology by itself is not enough, many companies have adopted a socio-technical approach in which the engineering/computing is technically well thought-through and integrated into the social and organisational aspect of the company via job and organisation redesign. Traditionally this has meant the development of more participative approaches to implementation strategy, usually involving "human factors" specialists.

There are many examples of this kind of approach which is well documented by Ettlie (1988). A good example can be seen in the loom shop application in the Westland Helicopters case study. The kind of shop floor involvement typified in this example, where operatives contributed local expertise to system design thus adding to what had typically been an exclusively engineering approach, shows the value of this approach. The limitations lie in the way the application is relating to the more strategic business questions.

The usual outcome for this strategy is quite positive in that the innovation is accepted and installed. However the impact on the business performance of the company in terms of profitability and competitiveness is not assured unless it also constitutes a strategic improvement in the manufacturing process. This strategy is strong on installation and implementation, but weak on exploitation of the new technology for business benefit.

4.   Low Concern for Technology and Business/High Concern for Organisation

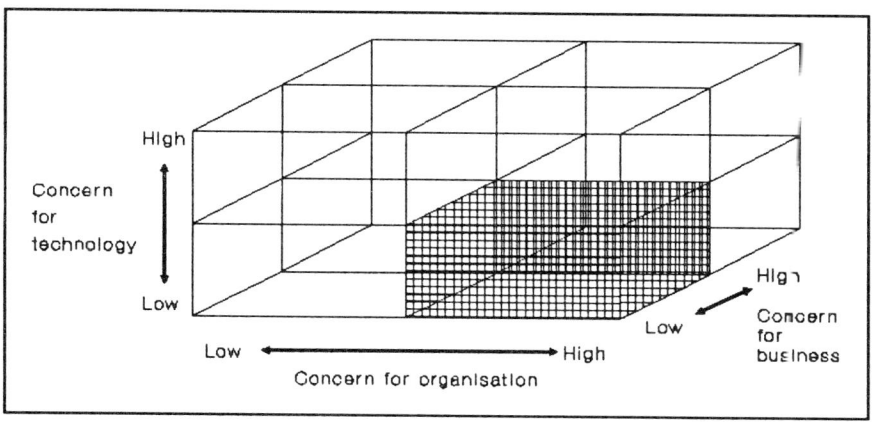

This strategy usually comprises a poorly organised and thought-through technical installation which is only marginally related to business requirement other than by the loosest, most superficial argument. However, the strength of the strategy lies in its ability to put people and organisation at the forefront of its thinking. The support, creativity and goodwill earned by this can sometimes compensate for the lack of technical excellence, and create a situation where the innovation can be 'muddled through' providing improved efficiency and effectiveness due to the social system changes. In this way, this particular strategy may be labelled a process-oriented strategy as it is more concerned with context and organisational infrastructure than with technical content. Traditionally this approach has been the province of organisation development consultants. Although rarely used with such a cyclopean focus in manufacturing companies, the significance of the organisational dimension can be seen in emerging approaches to manufacturing systems engineering such as those advocated by Parnaby 1988 and in the role played by organisational factors in the development of flexible infrastructures to support application systems in dynamic environments such as electronics, Childe 1989.

There is a variety of examples documented in the academic and trade literature which details the importance not only of redesigning and retooling manufacturing technology but also of reorganising manufacturing companies at both the strategic and job design levels. This reorganisation is not only concerned with the structural rearrangement of company organisation, but also with the cultural regeneration of the company. The suggestion that the success of Japanese manufacturing companies has more to do with organisational factors than technological excellence is well established, (for example see Schonberger 1986), with the main tools for achieving this being the Total Quality and Just-in-Time philosophies. When practised well, these philosophies are not just packaged techniques for company improvement, but ideologies which pervade company culture. Changing manufacturing companies using these can provide interventions which focus primarily on company organisation as the main driving force for regeneration.

5. High Concern for Technology and Business/Low Concern for Organisation

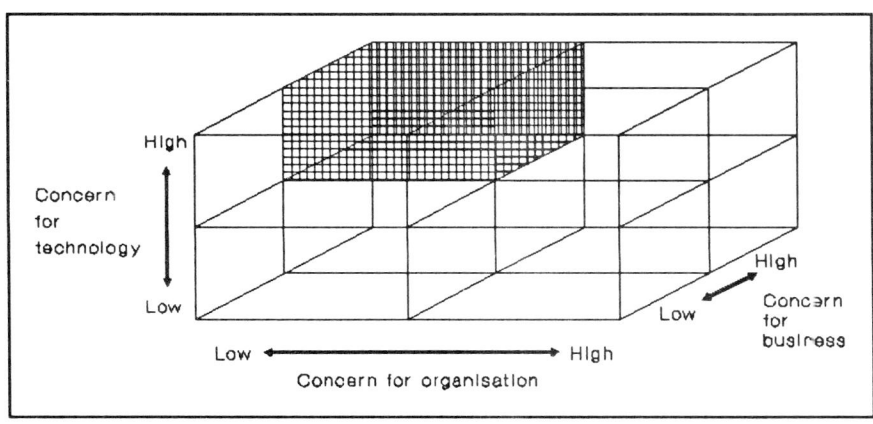

In this is strategy, a good business plan has led to a decision to purchase a particular technology, with a clear relationship established between the business plan and the ability of the technology to deliver. The technological installation strategy is well-defined and planned both internally and in relation to the management of external suppliers and vendors of equipment, and also may include external consultants. However, the lack of planning concerning an organisational change strategy can seriously inhibit the development of a final organisational design to support the change, and can create serious loss of goodwill among the members of the organisation. This may be manifest in IR problems specifically, or resistance to change generally, which will increase the timescales for the implementation.

An aerospace company manufacturing all kinds of aircraft bodyparts provided us with a good example of the strategy. The parent company was in the process of reorganising into business centres of excellence to meet an anticipated rise in demand for a particular European market. Hence the business case was of great concern to management although not well articulated below senior management level. The preoccupation with the technological change was central. A whole committee structure had developed to manage the development and introduction of six FMS.

This programme involved the conversion of existing Max-E-Trace machines with modifications to each costing some £700,000. To the modified machines it was proposed to add extras in terms of swarf extraction, automatic tool changing and automatic billet presentation. Already the facility had a significant number of machining centres in DNC mode linked by a LAN. On these machines, work scheduling and communications to production centre management, material stores, tool stores and maintenance was connected via a VAX to a mainframe used for factory management and engineering. Considerable in-house expertise had been developed in designing and building the FMSs which were to be linked into the existing communications system in the medium term, and expertise had been developed also by project managing such a significant change.

However, we found no evidence in the development of such a major innovation of any serious rethink of company organisation other than at the most gross structural level. Culturally the company remained untouched, and although jobs were redesigned as the new FMSs were brought on-stream, management "fell over" problems as they arose rather than had an overall design towards which to work and a clear implementation plan for getting there. The result was a dispute over the working of the new technology which focused on the two issues of "flexibility of working" and "continuous cover". What should have been the factory for the future became a new battleground for the past because of an insufficient implementation strategy which was based on two dimensions but ignored a third.

Overall it is true to say that this position on the cube occurs with more frequency than does any other which was found in our research.

6.   Low Concern for Technology and Organisation/High Concern for Business

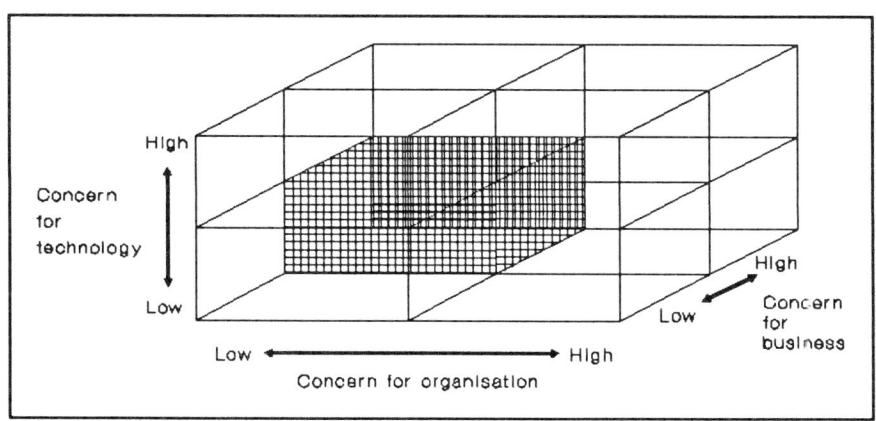

This strategy could be labelled the businessman's strategy. It comprises a well thought-out business strategy in which the competition is analysed, a competitive edge defined and a marketing and financial strategy decided which fits well. A manufacturing strategy may have been considered but, by and large, the delivery of an organisational form and a technological horizon to achieve the business strategy are undeveloped. The whole strategy therefore lacks concreteness and can fail because it leads to "analysis-paralysis". Unless resolved later in the whole process by buying in skills in the engineering and organisation design and development areas, the strategy becomes a good business idea which failed because of the inability to push it through into a real organisational and technical change.

7.   Low Concern for Technology/High Concern for Business and Organisation

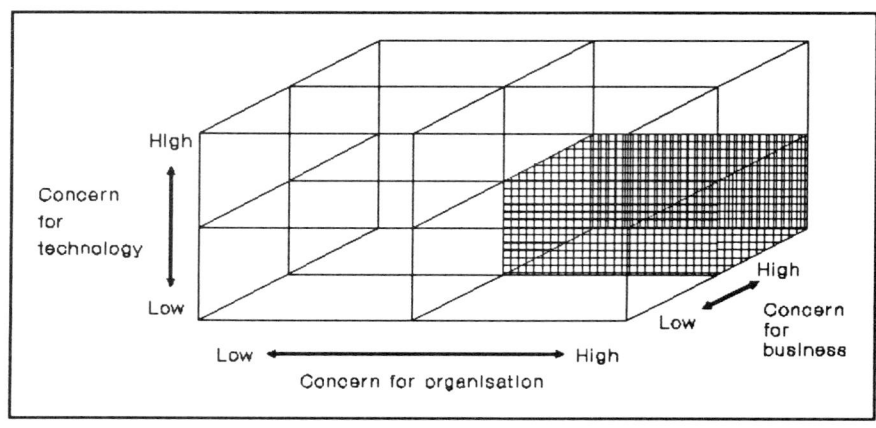

This is a non-technical approach and is unlikely in practice. It may characterise the early stages of planning for CIM implementation where companies have little or no technological capability or history. There is a clear business plan and strategy for organisational change, but engineering capability and the resulting technical installation strategy are sadly lacking. Such companies do have, however, a sound basis for proceeding, as long as engineering/computing expertise can be obtained from outside and be well-managed from within. Naturally such input leads to further revision of the business plan and organisational change strategy, but the final outcome is the equivalent of (8) below, albeit with a combination of internal and external resources.

It is possible to see how this strategy could be initiated in small- and medium-sized enterprises such as the case of Rotabroach outlined in Chapter 7. Although the Rotabroach management did not fall into the trap of having little concern for the technology they were using, small- and medium-sized companies often use relatively simple, well-tried and tested technology, or packages. Necessarily this can lead to a downgrading of the importance and significance of the technology itself. Rotabroach is a good example of way in which small and medium sized companies can use advanced manufacturing technology by purchasing simple, tried and tested machines to fulfil a clear business purpose. Allied to this needs to be a measured approach to implementation as evidenced in the Rotabroach case where the Managing Director's repeated exhortations for utilisation were significant in a smooth introduction and acceptance, coupled with an incremental (learning) approach to the introduction of each machine within an overall three year plan.

8.   High Concern for Technology/Business/Organisation

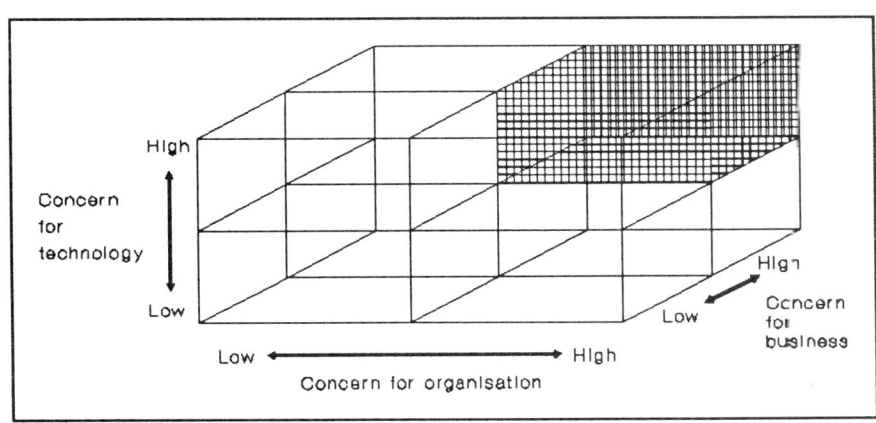

Clearly this is the preferred position of all eight potential positions within the implementation cube. All three dimensions are considered in some depth, and our experience would suggest that this is best done in the order business dimensions first, technological and organisational dimensions afterwards, thinking of each of the latter two as ways of delivering a pre-defined set of wider objectives. It is important to plan the technological and organisational dimensions in parallel for there may be aspects of the one which impact upon the other as the implementation proceeds. This raises the salience of adequate project planning and management if change is to come about in an orderly fashion on a wide variety of fronts.

Examples of this final approach to implementation are, in our experience, relatively few in achievement but are becoming available in increasing numbers. The two large case studies in this book concerning Westland Helicopters (see 'Out of Engineering and into Business' in Chapter 5) and Cummins Engine Company show strikingly what can be achieved with an integrated approach to change with equal concern for all three of the dimensions specified. Both of these examples were taken, interestingly enough, from brownfield sites where traditionally managements have suggested that most problems in introducing change would exist. Greenfield applications appear considerably easier. One company we investigated introduced a new flexible mechanical handling system into one of its sites. This site was greyfield in that it was geographically next to the original site. The company had already identified a need for decentralisation in order that its numerous markets could be best serviced by units targeted and organised precisely for this purpose. In this sense, clear business objectives were set prior to design and installation, and issues of overall organisation design had been addressed and instigated. The FMHS constituted a step function change for the manufacture of the particular product and was based on a model plant in Sweden which provided the 'vision' of the possible. Tonnage was increased with a staff reduction of approximately 60%. New working arrangements including flexible manning were approved. However, although this was a greyfield site, and the ease of introduction was facilitated by that, manning was taken from employees of the previous operation. This meant a major re-training and attitude change programme. Part of the latter was a function of recruitment, but was also achieved via a new industrial relations structure based on local plant bargaining and involving all four of the unions represented on the new site. This development has fed back into the company overall causing it to reflect on its industrial relations structure in total. A further, and in our opinion vital, ingredient to the change, was the introduction of a 'nine point plan' offering certain benefits but giving certain responsibilities to those who chose to sign it and work at the new site. Clearly, success here on what was a greenfield development on a brownfield base, was due to management paying equal attention to the three dimensions outlined above.

Concluding Comments

The above ideas, which are based on the approaches taken by many experienced companies in introducing advanced manufacturing systems, illustrate various implementation strategies and their relative strengths and weaknesses. The dimensions of the implementation cube are an attempt to capture the main sources of variety in these strategies. We have argued that success is determined by 'high concern' on all three dimensions of technology, organisation and business. In all the cases we examined, success could be traced to either the explicit or implicit management of these three aspects of implementation. The other feature critical to the successful introduction of advanced manufacturing systems, both in terms of hardware/software and new manufacturing methods, is the development of system wide implementation processes. The more integrated the advanced manufacturing system, the more a frame for implementation is required in terms of business and organisational dimensions. In our view, although all three dimensions are equally important, the business dimension is prepotent. It is the business dimension that represents the sine qua non of any successful change and it is the failure of some senior managements to recognise the radical impact that advanced manufacturing systems can have on the companies' capacity to compete in world markets that leads them to adopt inappropriate piecemeal implementation strategies which do not involve a fundamental reassessment of both manufacturing strategy and manufacturing organisation.

Whereas Chapter 2 has focused on the strategic orientations taken by various companies to implementation, Chapter 3 develops the ideas to the level of a practical methodology as an aid for management.

CHAPTER 3

An Outline for a Methodology[2]

Whereas Chapter 2 argued for a strategic view of implementation incorporating three dimensions, this chapter specifies a level of detail designed to make the concepts and ideas usable and relevant to practising managers. Arguably it has two main objectives.

First of all it identifies and discusses the key elements that are significant in implementing change associated with advanced manufacturing systems. Secondly it outlines an orderly set of procedures which comprise an implementation methodology for the management of the introduction of advanced manufacturing systems. The chapter ends with some specified ideas for the practical application of the methodology.

However, to start with, there is a brief discussion on the development of methodologies in general, in order to place subsequent ideas in context.

---

[2]This chapter is based on a paper given by the authors to the inaugural conference of the British Academy of Management, University of Warwick 1987, and later published in part as 'Managing Rapid Change', *Management Decision*, 20,1, Spring 1988, MCB University Press.

Developing a Methodology

It is clear from everyday experience such as following recipes to make cakes or assemble DIY furniture, that it is impossible to cover every aspect in a guide to practical action, such that success is always guaranteed! Implementation of new technology in manufacturing is a vastly more complex act than baking a cake or assembling a wardrobe, so it is clear that any recipe for action is forced to be limited in scope and inevitably relies on the inherent intelligence, experience, skills, etc. of the implementers in practice.

In examining everyday experience it is apparent that the most useful feature of methodologies or recipes is that they provide a conceptual map to guide practice. The useful map contains sufficient detail to inform the practitioner what to do and how to do it, but it inevitably takes some degree of taken for granted knowledge and skill. For example, most recipes do not inform you how to crack an egg or drill a hole. In fact, the most useful recipes for action focus on the tricky, more difficult bits and skirt over the parts where people have sufficient experience to be successful without detailed guidance.

The other basic feature of recipes worth noting is that they focus on what and how. They inform the user about what needs to be attended to and how this should be achieved. The DIY wardrobe recipe for action usually contains a comprehensive list of parts and a method for transforming them into a wardrobe. The cake recipe contains a list of ingredients and a method for mixing them and transforming them into a cake.

We decided to take these observations from everyday experiences and use them to shape the design specification for a useful implementation methodology. Our aim was to make this methodology have the following characteristics:

    1.   The methodology should focus on those aspects of implementation which are

most 'tricky', i.e. out of the ordinary compared with other experiences of implementing technology.

2.   The methodology should provide a conceptual framework for thinking about and making sense of the implementation process, incorporating the 'best' features from the implicit and explicit methods used by the companies studied.

3.   The methodology should highlight the issues that need to be attended to in implementation. It should have the capacity to address a wide range of technological, managerial and organisation dimensions.

4.   The methodology should suggest how the transformation (i.e. implementation) can be achieved. It should be a methodology for achieving morphogenic change.

Elements of a Methodology

From our work we have been able to distil nine elements which are key to a successful implementation methodology. Not all of the original applications which we studied exhibited all of these elements, but each of the elements recurred sufficiently regularly to justify their inclusion in this overall list. Sometimes respondents referred to these elements as serious omissions in the implementation process. Specific case examples can be found in later chapters. The final list of key 'ingredients' was as follows:-

(i)   **Business driven** - the one and only reason for investing was to improve the competitiveness of the company. Any other reason was peripheral to this. Particularly we identified a variety of superficial reasons such as:

'Because ... our competitors had one,

...our engineers insisted we needed one,

...a senior manager desired one',

and others were heard, sadly, all too often. We came to the conclusion that the only sound reason was a comprehensive business (not just financial) justification made on the basis of a comprehensive audit of the company situation. This ensures that any investment and change programme is related directly to specific, attainable and possibly measurable objectives. This is achieved by (ii) below.

(ii) **Back-to-basics rethink** - the implementation of new technology usually involves step-function (morphogenic) change rather than incremental (morphostatic) change. Step-function change requires primarily a fundamental analysis of the business situation and reappraisal of business objectives along various dimensions of competitiveness. A technological horizon and an organisation design or redesign are important elements of the future vision which has to be built as part of the rethinking process.

(iii) **Top management driven** - our own research agrees with many others that the best way to achieve morphogenic change is top down. Usually this involves at least one senior management 'champion' in the first instance, but needs to receive much wider senior management support if it is to be successful. However, we would add to the existing views by identifying a need to ensure a strong 'bottom-up' involvement. Whilst the process is led, or driven, from the top, the messages of the required change have to be cascaded throughout the management organisation in order to develop the 'corporate will' to change, and a dialogue achieved. This is further explored in (vii) below.

(iv)   **Front end-back end** - business and managerial issues are best considered at the 'front end' of the change process, rather than being the tail on the technological dog. Implicitly or explicitly, successful implementers of change worked to a model which emphasised first the specification of a business and management strategy, secondly the ascertainment of a manufacturing strategy, and then only at the back end of the process, the development of a technological horizon capable of delivering. Sometimes, in following such a methodology, companies had reversed the decision to invest, whilst other examples showed significant amendments from original plans.

(v)   **Integrated change strategies** - successful implementation involved changing the company on a wide variety of fronts. This meant an overall change in terms of competitiveness, but also, technologically, structurally, and culturally. The aim of developing an integrated change strategy was to get the company 'firing on all cylinders at once'. To conceive the problem solely in technological terms provided far too limiting a view, and tended to achieve morphostatic or replacement objectives rather than the required morphogenisis.

(vi)   **Investment in people as well as technology** - the knock-on effect of implementing new technology into the social organisation of companies requires considerable investment. Whereas investment in AMT provides a technological re-tooling, so new skills and just as significantly, new attitudes, are required within companies. This is true for all employees but particularly true for the whole of the management structure where there are needs for greatly increased flexibility and change. Clearly education and training have a major role to play in bringing about this 'social re-tooling'.

(vii)   **Everybody on board** - any effective implementation must be cascaded down throughout the organisation, particularly the management hierarchy. Without this, morphogenesis cannot be achieved and new technology is implemented into a situation of traditional structures, attitudes and work practices. It is in this situation that our evidence suggests it is most likely to fail. Developing the corporate will to change by resourcing and facilitating a cascading structure of workshops educating management of whatever level about the need for change, and encouraging participation in the development of the design for change and particularly feeding back the local impact of suggested changes, is a vital part of the process of effective implementation.

(viii) **Clear targets** - whereas (vii) above emphasises the generation of commitment via a cascade of ideas down the company, it is equally important to restructure where necessary. Often this involves the breaking down of old barriers and the establishing of new departments, for new technology often does not respect traditional structures. Usually there is an increased emphasis on integration and working together which is in stark contrast to the more highly differentiated forms of traditional manufacturing. The most important step in this restructuring process is the generation of clear targets for managers within the new structure.

(ix) **Time scales** - both short and long term are important, particularly as change is brought about in step-function revolutions followed by periods of relative stability. More is said about this below.

Whilst all of these elements were important, we were surprised to hear the emphasis placed by engineers as well as managers on the importance of culture change as being the key issue for successful implementation. Increasingly we have come to view culture as the shared, taken-for-granted (TFG) assumptions of the members of the company. For example, companies who had introduced Just in Time production methodologies reported the importance of changing TFGs embedded in the previous batch scheduling philosophies, and those who had introduced Total Quality reported the same need to change the orientation of those born and raised in a quality control/quality assurance environment.

Culture is a difficult concept to grasp when defined as shared TFG assumptions, and a number of points are worth noting. Whilst on the one hand it is the shared TFGs that make social and organisational life possible, on the other it is these same TFGs that provide structure and security and hence inhibit change and adaptability. Often they are implicit and can only be inferred from behaviour, being rarely up for scrutiny, and making overt appearances only when some infringement occurs. However, they do tend to permeate the whole of organisational life, giving a sense of overall purpose, and enabling members of the organisation to identify with the whole, as well as being a source of control over individual behaviour. Successful implementation, from our experience of manufacturing, requires that any implementation methodology addresses these company-wide issues prior to installation, for changes in these TFGs, such as changes in attitudes to quality, costs, productivity, lead times, customer service, are key to institutionalising morphogenic change.

Whilst the above gives a list of 'ingredients', it does not specify the process whereby they can be organised to produce the desired outcome. This is the job of the next section.

An Outline Methodology

Our main aim in doing this work was to produce a form and scope of a practical methodology for implementation. Our initial orientation had been to produce a detailed methodology probably related to project management. However, our experiences in researching the companies involved led us time and again toward the conclusion that the development of a more strategic methodology than mere project management was needed. This was for two reasons.

Firstly, strategic variables were of prime concern to management and it was this strategic impact which made implementation of the technology different from other forms of technological innovation. Secondly, successful implementation seemed to be governed by whether it was seen, implicitly or explicitly, as part of the strategic response of the company, i.e. central to business performance. Strategy was certainly the 'tricky' bit that our methodology needed to address in the first instance.

For both of these reasons, we decided to concentrate on a strategic view of the implementation problem. In the event, only one of the companies we studied had adopted a formal methodology to help them manage implementation, and essentially this was a project management methodology for ensuring agreement by the wide-ranging group of people involved in the implementation. However, all of the other companies had evolved implicit methodologies on the basis of their previous experience of implementation, and their view of the particular needs of the new technology that they were introducing. Quickly it became clear from the managers involved that, whilst they were enthused by the approach taken by themselves, there was a great need for a systematic methodology.

Our methodology begins with an audit of the current company business situation and manufacturing strategy, which incorporates a clear assessment of what is going well, what opportunities there are, and where problem areas exist. Sometimes this audit is the response to a crisis, sometimes the response to new knowledge identified with a 'champion' of a particular cause. Inevitably, this audit involves the exploration of various alternatives along a variety of competitive dimensions (cost, productivity, quality, lead time, etc.) and forms the basis for the development of manufacturing strategy.

Developing a manufacturing strategy frequently demands that senior manufacturing managers fundamentally rethink their approach to manufacture, considering the kinds of questions outlined earlier.

In this way, the methodology begins with ideas and concepts about business and manufacturing strategy in relation to environmental opportunities, pressures and constraints (Fig.3.1).

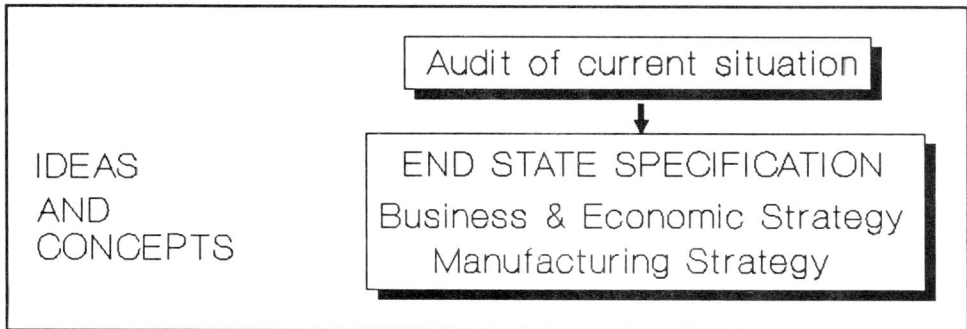

Figure 3.1

Once top management has agreed the current situation and a way forward, the methodology moves into a design for delivery phase in both organisational and technological terms. On the organisation side, the methodology involves restructuring and culture change, both designed to bring about an organisational delivery system to exploit the new technology to the full and contribute to the agreed end state. At the same time, production managers, engineers and computer scientists need to be designing a technological horizon to deliver the manufacturing strategy (Fig. 3.2).

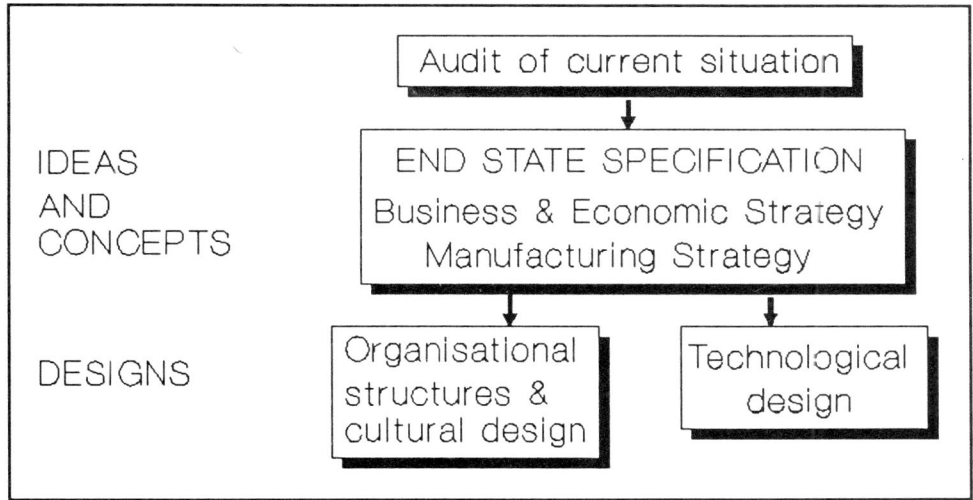

Figure 3.2

Finally, the methodology is delivered into the company by the development of three activities, the objective of which is to deliver the designs outlined earlier. These three are outlined in Figure 3.3 and involve on the one side the development of an approach to technological installation, and, on the other, to structural and cultural change.

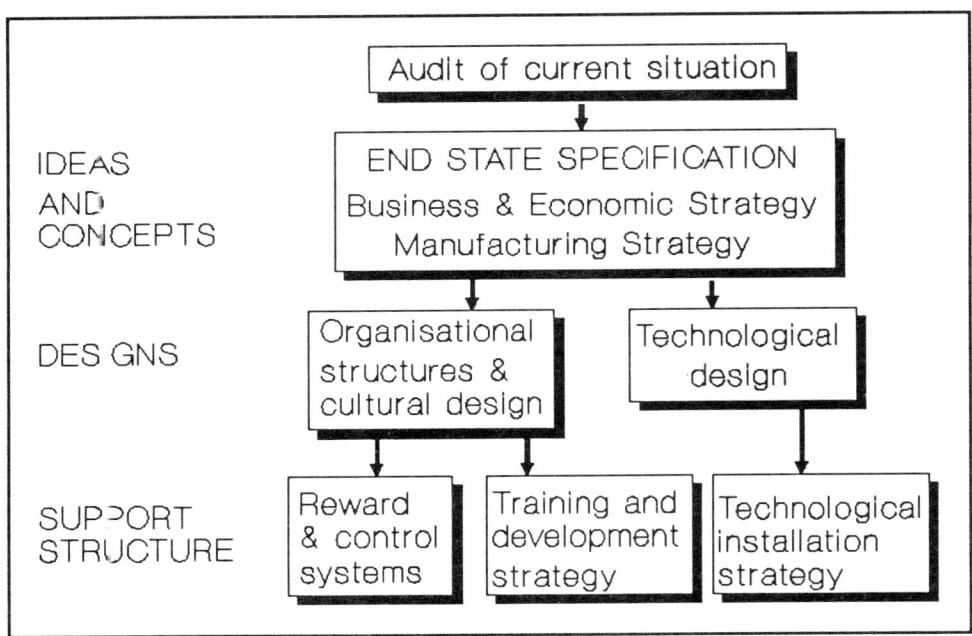

Figure 3.3

These are brought about by changing reward and control systems, and by devising a training and development strategy to support the required structural and cultural changes. Nowhere is the importance of the latter better illustrated than in the implementation of MRP II, where attitude change as well as understanding is vital if the new system is to avoid problems of data capture.

Making the Methodology Practical for Managerial Use

Whilst the methodology outlined above provides a roadmap for implementation, it does not identify the key managerial activities involved, nor does it mention timescales and the pace of change. From our work we have been able to identify the following two key managerial activities in implementation which enable the list of 'ingredients', to be organised in such a way as to produce the desired outcome. Companies implementing AMT effectively have succeeded by paying particular attention to the pacing of the whole change endeavour. They have concentrated on two particular aspects of the pace of change which we have labelled firstly 'the sprint' and secondly 'the performance ratchet'.

The Sprint

Successful implementation and successful regeneration seems best handled by the initiation in the first instance of a short-term process of transformational change. For companies that have languished for years in the downward cycle of declining profitability, increasing cost, etc., a sprint which is planned and executed on a wide variety of fronts, provides a strong attraction as a way forward.

Sprints have a number of distinct characteristics which are vital if they are to succeed:

a)   They are initiated, developed and managed right from the very top of the company. The levels of capital expenditure often found in a sprint, coupled with the structural reorganisation of the company (perhaps including redundancy programmes) require that the whole process is top management led. Without the support of the chief executive and several key senior managers around him, the sprint will not succeed. They are usually initiated by a senior management audit (see below) of the company situation.

b)   Sprints are focused on one of the few key dimensions of competitive performance central to the survival and development of the company, e.g. cost reduction or maintenance, quality improvement, inventory turnover increase etc. These dimensions should contain quantifiable targets for performance improvement and be clearly communicated throughout the management of the company, along with the central or core values critical to the achievement of these targets (see cascade below).

c) They involve all members of the company and are seen to do so, but particularly all members of management must be involved and committed to the idea of the sprint and the key dimensions of performance improvement.

d) They have a fixed timescale in which the improvement must be made.

Sprints often begin with senior management, or even the chief executive alone, auditing the competitive position of the company and clearly articulating a vision toward which the company should be driving. To be effective, this auditing and vision-building process has to include a thorough analysis of the competitive position of the company using whatever dimensions are salient, a ruthless assessment of current policies, performance, and resources, as well as a statement of required changes and a method for bringing them about. This analysis and vision-building activity may use tools such as BCG analysis or stakeholder analysis. In any event, the main aim is to identify both a business plan and a set of central values (no more than four to six) with which company members can identify. In this way, sprints contain *auditing and vision-building* activities which are the concern of the top team and provide the context in which sprints can be driven.

Involvement of all levels of management is best handled by a *cascade* of the ideas generated by the top team. Once the initial work has been done and agreed by the top management, then the 'message' (both business plan and central values) can be cascaded through the management team. This may be done after structural reorganisation, and will usually involve the use of training and development as well as change in reward and control systems. A cascade is usually best handled by a series of workshops in which senior managers present a view of the future, and workshop delegates are invited to comment, discuss, and look at the implications for their own areas of work, as well as feedback upon the ideas given to them. This cascade process needs to involve the whole management structure and may consume large quantities of resource. In fact our investigations revealed that as a rule of thumb, the programme to initiate and introduce the change into the company organisation requires resourcing to the level of approximately 10% of the total annual labour cost of the company.

Therefore, cascades provide an opportunity for the management of the company to receive, either verbally, in writing, or both, the thoughts of the top team, and to explore and respond to these ideas. Cascades often contain an educational input, particularly where new technology is concerned, but it is not sufficient to regard a cascade only as a 'tell' situation. Often they are best managed as a series of in-company workshops in which senior managers can present audit/vision-building ideas and have them explored and discussed, and receive feedback on the local difficulties that strategic decisions may generate. A cascade is nothing if not a method of building commitment throughout the management team to a particular view of the company situation and a way forward toward a strategic vision. It is because of this that to regard a cascade as nothing other than an educational input can lead to problems in controlling implementation later in the process.

Setting up a sprint is often the first company-wide activity in managing strategic change inside the company. It is an activity which usually, after the early shock, engenders much support and enthusiasm, but which consumes much resource. At a detailed level, sprints change not only the company strategy, but also structural relations, job designs and reward and control systems. Often they focus upon large capital investments. However, the most significant effect reported to us was that sprints were a method of initiating culture change within companies. They did this by questioning the very taken-for-granted assumptions made by management in the company, and forcing a rethinking of work practice. Whereas it was easy by comparison to install a new technical specification into companies, this 'organisational engineering/social re-tooling' to match future requirements was a much more difficult process.

Whilst sprints with their auditing, vision-building and cascading activities are vital for initiating regeneration and driving it forward, they are not sufficient if change is to be sustained in the medium and long term. Indeed, it would not be desirable for companies to continue to change at rapid rates usually associated with a sprint. Therefore, another mechanism is required, if companies are not to slip back into stagnation and crisis once the initial burst of energy created by the sprint is dissipated. We have called this other mechanism 'the performance ratchet'.

The Performance Ratchet

The aim of the sprint may be to bring an ailing company back 'up to speed' and to make it competitive once more, or to enable a strong company to become a world-class performer. As was argued above, it is a prime mechanism for bringing about transformational or rapid change in companies. If sprinting contains the idea of transformational change, then the performance ratchet contains the idea of incremental change.

The energy, commitment and enthusiasm demanded from management and organisation members is difficult to sustain over a protracted period. Indeed, the rate of change during the sprint would itself prove a major stumbling block to organisational effectiveness were it to continue unabated for a great length of time. Once the company is launched on its new path, with the accompanying technological advance, a further mechanism is required to ensure that the technology is exploited to maximum business effect. The performance ratchet does this by ensuring that minor incremental (morphostatic) changes are made by those most appropriate to initiate and support them.

Performance ratchets are best explained by observing them in practice. They usually take the form of senior management requesting from operating divisions, departmental managers or section leaders , the kind of performance improvement that they (the local management) could reasonably expect to make over the next time period. Local management are asked to offer a range of performance improvement targets, coupled with an estimate of the certainty by which these performance improvements could be made. These statements and estimates are brought together by senior management and are approved or not depending on how they fit together into an overall business plan. Performance ratchets are not to be confused with the operating plan of the company or regarded purely as management control devices. They are a way of continuing to get the company to change and to plan changes on an incremental basis. In this sense, they address issues which are out of the ordinary in terms of business planning, and which constitute small drives toward managing incremental change on a limited front over a certain planning period. For example, a division, department or section may get the go-ahead to drive toward a 20% quality improvement over the next six months if this is seen as vital in delivering to a particular customer or market segment. In a different organisation inventory turn may be targeted to double over the next year. Quality improvement is particularly significant as an area of continuous improvement in manufacturing, but we have found examples where safety improvement can be targeted where this is central to the survival of the company.

Performance ratchets organised in this way have the benefit of continuing the idea of change initiated by the sprint. They provide a link between strategic direction and operations during periods of incremental change within the company, and can operate on a series of 'themes' as far as company members are concerned, giving a clear but limited focus on which to direct activities. Performance ratchets constitute a means by which continuous performance improvement can be made, and the focus on change initiated by the sprint is not allowed to fade away.

Concluding Comments

The methodology outlines the strategic orientation required to establish a direction and pace for technological change in the manufacturing sector. It specifies the form and scope of an implementation methodology and sets out the strategic dimensions important for success. Particularly, it focuses on the need for a revolutionary orientation on a wide yet integrated set of fronts, which emphasise the development of a management and manufacturing strategy, and an organisation redesign, as variables of equal importance to technological installation.

A feature of strategic methodologies, however, is that by definition they tend to be non-specific. In many ways, they resemble the derivation of an equation specifying relations between variables without assigning values to them. This is the job of individual managements, perhaps with consultancy help, who must specify the detail of their situation and apply the methodology to their company and technology.

The methodology outlines an approach which has proved successful when allied to far-sighted and confident management. Particularly the methodology requires that managers using it understand the difference between leadership and management. Whereas the latter is usually associated with control and administration, and at best morphostatic change, leadership is closely associated with formulating the vision and generating the corporate will throughout the enterprise to achieve that vision. These latter activities place great stress on those involved. This methodology outlines a strategic tool to get the job done.

Having outlined the means for obtaining change in Chapter 3, Chapter 4 goes on to specify the organisational design ends toward which companies should be working to successfully adopt, and optimally exploit new technology and production methods.

CHAPTER 4

Manufacturing Organisation for the 1990s[3]

Whereas this book has concentrated so far on the issue of effectively implementing and exploiting AMT, this chapter now turns attention to the design of an appropriate company organisation to support the technology. Whilst it is necessary and important to initiate a 'sprint' and establish a 'performance ratchet' if change is to be introduced effectively, so far no guidelines have been outlined itemising the emerging principles of organisation which have proved to be useful by lead-edge companies. Such guidelines can be thought of as providing an outline template of manufacturing organisation which both generates ideas for manufacturing managers as well as giving an indication of an 'ideal type' against which managements can assess their current arrangements. In short, this chapter seeks to answer the question 'sprint toward what sort of organisation?'

The chapter is in three main sections. Firstly it outlines the overall changes which are occurring in the manufacturing environment and argues that currently we are experiencing a step function shift in manufacturing technology equivalent to the transformations brought

---

[3]This chapter is based on a series of papers by the authors together with J.Bessant, P.Levy and C.Ley - see references.

about by the introduction of steam power or electricity. An understanding of this context is important for it is this development from mass production technology to information and communications technology which is requiring a major rethink of organisational form if it is to be exploited to achieve competitive edge. Secondly it explores the detailed impact of integrated technologies at the level of the firm. It does this primarily by focusing the discussion on ideas drawn from the work of Henry Mintzberg (1979, 1983, 1989). In this discussion the impact and opportunities offered by integrated technologies to modify both the structure and the ways in which the company organisation is coordinated are explored in some depth. Lastly the chapter will outline detailed changes that can be observed at the level of work and management organisation as well as the relationship between organisations, which will give managers specific clues when generating design alternatives.

This in concluding chapter of
Bring all other chapters together

Innovation in Organisation Design

Ettlie (1988) suggests that the most successful approach to implementing new technology is 'synchronous innovation', that is, to innovate simultaneously along both technical and organisational dimensions (see Fig. 4.1). He argues that it is important not to let the organisation be determined solely by the technology. A key feature of the current generation of integrated AMTs is that they are what Fleck (1987) terms 'configurational technologies'. That is, they affect the whole way manufacturing is configured, not just the technology itself. In Chapter 3 we argued that the most successful approach to implementing AMT looked at the three dimensions of concerns for technology, organisation and business together and that exploitation demanded configuring manufacturing systems which best delivered competitive edge through integrated technical and organisational arrangements. Therefore there seems to be some considerable support for the need to reconfigure both company organisation as well as business purpose when investing in integrated systems. But why should this be? One reason was explored in Chapter 1, where an alternative model of transformational or morphogenic change was suggested as a solution to the incremental or morphostatic models usually adopted implicitly by managers making changes.

50

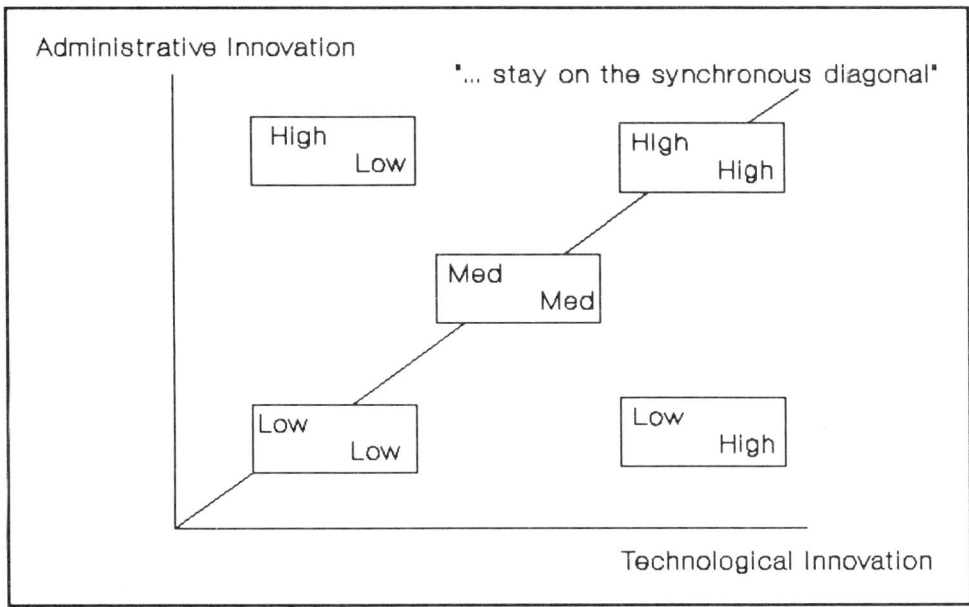

Figure 4.1 Synchronous innovation

Another reason which has been suggested is the notion that one important feature of the present wave of changes is that they are significantly more radical than those which have characterised the pattern of economic and technological development over the past forty years or so. There is significant difference along two dimensions. First the environment for manufacturing since the mid-1970s has been characterised by a shift towards massively increased competition, globalization of markets, fragmenting demand and growing emphasis on non-price factors. At the same time the technological thrust has changed, with an emphasis on *integration* of systems rather than discrete substitution of existing systems.

In looking at the changes that are taking place in manufacturing we have found the concept of 'paradigm shift' useful. Kuhn (1962) has argued that change in science proceeds by a combination of small incremental developments punctuated by periodic 'paradigm shifts' in which the whole structure of how scientists see the world is altered - as for example, in the transition form Newtonian physics to quantum physics. Such a shift brings about enormous opportunities for new thought and sets the dominant pattern for the next period of incremental development in science.

Our view, following Perez and Freeman (1988), is that it is possible to argue that the present trends in technology represent part of a paradigm shift in manufacturing. In considering so-called long-waves in economic development associated with major technological innovations (such as steam power or electricity) they introduce the idea of a 'techno-economic paradigm'.

Dosi (1982) extended the idea of paradigms to the field of technological innovation, arguing that change in this field also proceeds along certain trajectories which are defined by the currently dominant paradigm, and Perez (1984) took this further using the idea of a 'techno-economic paradigm' which defines the 'common sense' rules which govern the workings - structural and technical - of industrial society and set the pattern of best practice. These persist for extended periods of time but eventually become increasingly inappropriate - just as, in Kuhn's scientific paradigms, particular ways of seeing become increasingly limited in their ability to explain new observations and experimental results. Eventually the paradigm shifts and a new one emerges - and the cycle repeats itself. We can represent this process as a series of 'S' curves, and our current position as jumping from the fourth to the fifth curve, as shown in Figure 4.2

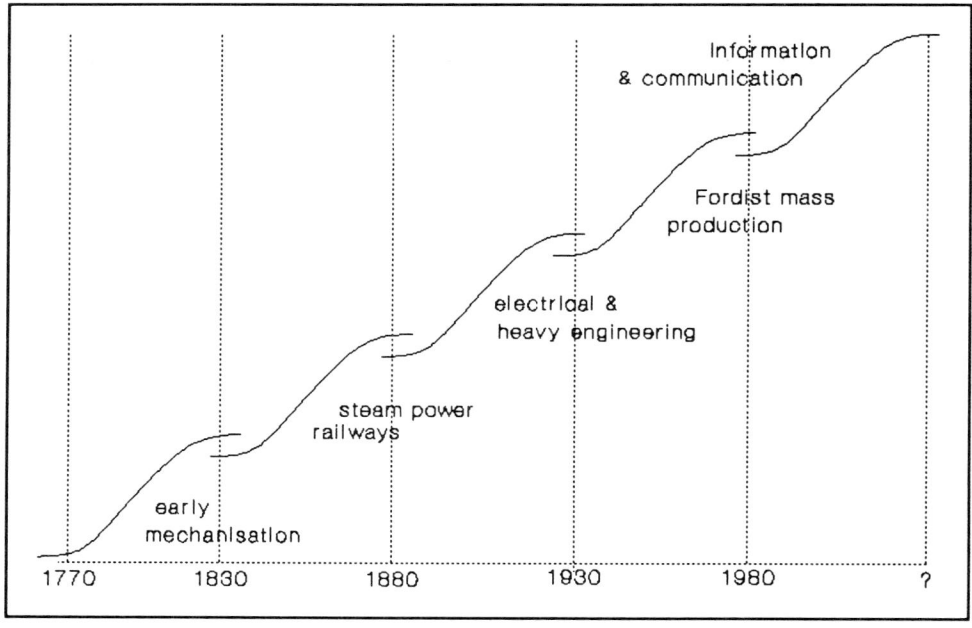

**Figure 4.2 Technological performance
and economic change**

Having set up such a model, we can interpret many of the features of late 20th century industrial society in terms of such a shift in paradigms. For example, there is growing agreement that the mass production models of production and consumption which dominated the first half of this century are giving way to more fragmented patterns of demand and more flexible modes of production. Such models are incompatible with older forms of organisation, especially those stressing division of labour and rigid bureaucratic organisational forms. Instead we are seeing the emergence of alternative, more flexible arrangements based on networking and decentralisation.

At the same time, industries such as computers and industrial automation, emerging during the 1950s and 60s have now matured to the point where information technology offers a pervasive, low-cost resource which has extremely wide potential applicability.

Discussions about organisation design tend to follow a framework based on 'contingency theory' which argues that there is no single 'best' structure but rather that firms seek to get the best fit between their structure and their particular operating and environmental contingencies. The role of technology in this is widely acknowledged, and it is clear that radical changes in both technology and the environment in which firms operate are likely to lead to similarly dramatic shifts in the pattern of organisation structure and process. The problem is that, whilst it is possible to see the increasing inappropriateness of the old paradigm (with its implied 'best practice' modes for organisation design), there is no blueprint for the one which is newly emerging. Instead we are at the early stage where there is only speculation and experiment, often with competing models existing alongside those trying to keep the old model working. Following this argument, organisational form can be seen as being in transition in many companies, which can be viewed as undertaking 'natural experiments' in attempting to establish best practice in reconfiguring.

It is interesting to note that in Kuhn's original work on paradigms in his book 'The Structure of Scientific Revolutions', he observed, that although paradigm shifts represent a morphogenic/transformational change in outlook and thinking about a particular area of science, these changes did not take place in this fashion in the scientific community involved in that area of scientific enquiry. He noted that although an existing paradigm may be shown to have shortcomings in its ability to cope with the phenomena it purports to illuminate, it is retained and upheld by most practitioners until the time when an alternative paradigm is articulated and adopted by influential members of the practitioner group. Kuhn observes that there was a long time lapse between the first publication of quantum theories in physics and the wholesale adoption of this paradigm in favour of Newtonian mechanics. What appears to happen during periods of transition is that when paradigm practitioners find that their existing knowledge and assumptions are failing to produce adequate solutions, some of them seek alternatives, either elsewhere or from first principles - but the vast majority persist with existing recipes. In the case of the mass production paradigm some have looked to Japan for inspiration, others have actively experimented with novel organisational forms in an attempt to develop organisations more appropriate to current market and technological conditions.

Although the development of integrated automation has been going on for some twenty years it is only recently that the level of diffusion has begun to accelerate. Correspondingly, organisational adaptation is still a relatively isolated phenomenon and the above patterns are by no means clearly established in all firms. Nevertheless there seems to be growing acceptance that what is happening in the economic and technological environment does represent a major shift rather than a 'more of the same' pattern of evolution, with the consequence that firms will increasingly need to explore ways of adapting their organisational structures and processes. In this process of adaptation and development the contribution from both experience and theory will be important. The problem is that much of the present body of knowledge concerning technology and organisation design stems from research carried out within the old paradigm - for example, much of it is based on discrete, stand-alone technologies rather than integrated systems. A further complication comes from the fact that the new technologies themselves are by no means mature and established. Unlike earlier generations where some form of technological determinism could be identified, current generations are characterised by a high level of potential choice about their design and organisational implementation. (Sorge et al, 1982, Boddy and Buchanan, 1984). As was argued earlier, they are essentially 'configurational' and subject to a relatively high degree of organisational and social shaping (Fleck, 1987).

The result is that it is impossible to predict in any detailed way what the emerging blueprint for the new paradigm for manufacturing organisation will look like. That said, we believe that some clues can be found in the development of existing theoretical models dealing with organisational structure and design, as well as the documenting of in-house reorganisation by lead-edge companies providing 'natural experiments' in establishing parameters for the new paradigm of manufacturing organisation.

Theoretical Predictions

If theory is to be more useful to managers and other practitioners involved in implementing and using computer-integrated technologies, it needs to be able to move away from general to specific predictions about the nature of organisational arrangements which best support their competitive exploitation. Mintzberg (1979) offers a well-developed framework for understanding organisational design. We have found this useful in beginning to uncover some of the ways in which the configuration of the basic ways in which organisations are coordinated are affected in companies that are using computer-integrated technologies.

Mintzberg suggests that organisation structure can be defined as,

> 'the sum total of the ways in which its labour is divided into distinct tasks and then its coordination is achieved among these tasks'.

This is very similar to Lawrence and Lorsch's (1967) emphasis on 'differentiation - integration' as a key dimension underpinning organisational design. Differentation is the way the task is split up and integration is the means of gluing them together again. Mintzberg sees a limited number of basic coordinating mechanisms which constitute the 'glue' that holds organisations together. They are the means by which the tasks of individuals and groups are coordinated in the achievement of the organisation's overall mission and purpose. He originally enumerated five basic coordination mechanisms. Apart from mutual adjustment and direct supervision, he outlined three forms of standardisation which coordinated activity. These were, standardisation of work processes, standardisation of skills and standardisation of work outputs. In later works (Mintzberg, 1983, 1989) he describes a another form of standardisation, that of standardisation of ideology (beliefs and values) which coordinates activities through shared purpose rather than via standardised procedures or work processes or outputs. Each of his original mechanisms can be seen diagrammatically in the figures below, the importance of the sixth coordination mechanism, ideology, is explored below (see page 62).

Mutual adjustment achieves the coordination of work by the simple process of informal communication between those involved in doing the work. Under mutual adjustment control of work rests in the hands of the doers. Its strength is that it facilitates fast, flexible and responsive adaptation to work demands as they arise in situations where no one person is capable of understanding all the inputs required in the time available. Its weakness is that it needs

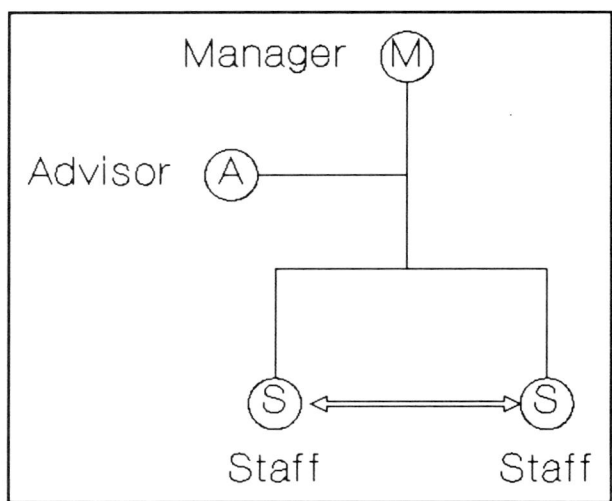

Figure 4.3 Mutual Adjustment

those who are to mutually adjust to be able to communicate informally, that is usually in direct physical proximity. Kanban cells and autonomous work groups are organisational arrangements which are coordinated mainly through mutual adjustment. In autonomous work groups multi-skilled workers manage a defined set of tasks between themselves by rotating tasks between workers and by flexibly adjusting to demands as they arise. Kanban cells are designed to enable flexible adjustment between work stations in order to smooth work flow.

Direct supervision achieves coordination by having one person take responsibility for the work of others, issuing instructions to them and monitoring their actions. In effect one brain coordinates several hands. Direct supervision is suited to situations where the work is simple, leisurely, or predictable enough for one person to cope in coordinating the work of others within the required time. Direct supervision is what many people think of when they talk about organisation structure, for it is the coordinating relationship defined in tree hierarchies. It is the form of coordination activity with which most of us are most familiar, and it is certainly the most dominant form of coordination mechanism under the old Fordist mass production paradigm of manufacturing organisation. Under those manufacturing arrangements

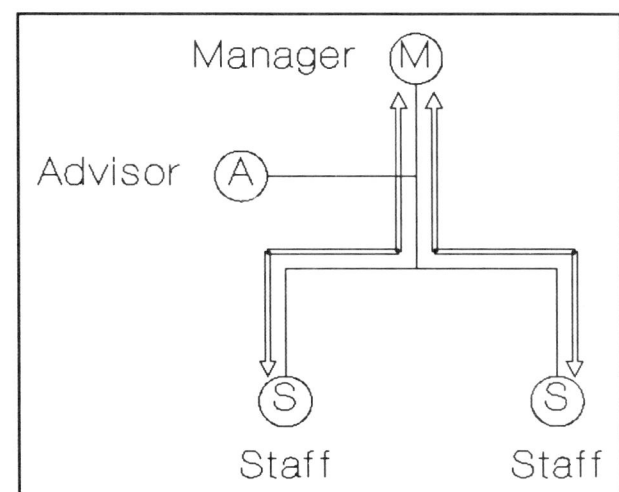

Figure 4.4 Direct Supervision

long hierarchies had to be established, the prime function of which was to coordinate the activities of those within them. Moreover, the roles of the managers under this coordination method became focused on control through bureaucracy. The transformation in manufacturing organisation that has to take place if we are to become competitive again has much to do with de-emphasising the importance of direct supervsion as a prime coordinating mechanism in manufacturing companies.

Standardisation of work processes occurs when the contents of the work are specified or programmed. This coordination mechanism is well suited to work processes which are predictable and therefore it is possible to specify how they should be carried out. Traditional mass production assembly lines epitomise work arrangements which are largely coordinated through standardisation of work processes. Work study practitioners define their distinctive expertise in developing these standards.

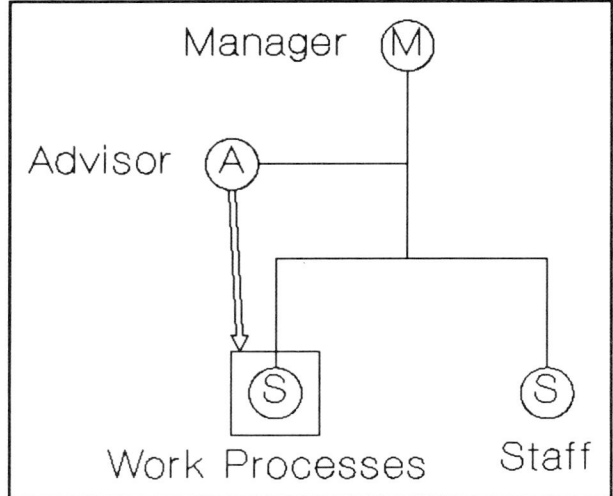

Figure 4.5 Standardisation of Work Processes

Standardisation of skills and knowledge coordinates work by controlling inputs. The role of training, and in particular professional training, is to facilitate coordination by standardising how work is approached and done. Although the skilled person can work autonomously, coordinating their input with others, there is a sense in which their training leads to a programmed/standardised performance. Skilled people are taught how to see and think about their work, what counts as important, what methods/techniques to employ, how to relate to their workers etc., etc. All manufacturing organisations rely on standardisation of inputs to coordinate some of the work of

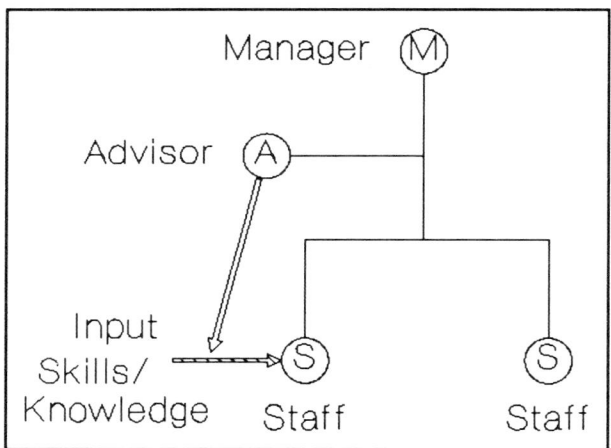

Figure 4.6 Standardisation of Skills & Knowledge

professionals and skilled staff. Design engineers are able to work together in designing a complex new product based on their common training which implicitly structures the way they approach the design task and underpins many of their shared assumptions.

Standardisation of Work Outputs is where the results of work are specified. This form of coordination is unconcerned with how work is done, only that the results comply with performance standards. The strengths of this coordination mechanism are that staff have high autonomy provided they meet the required level of performance. This form of coordination is familiar in most organisations and piece-work is an example of its direct use.

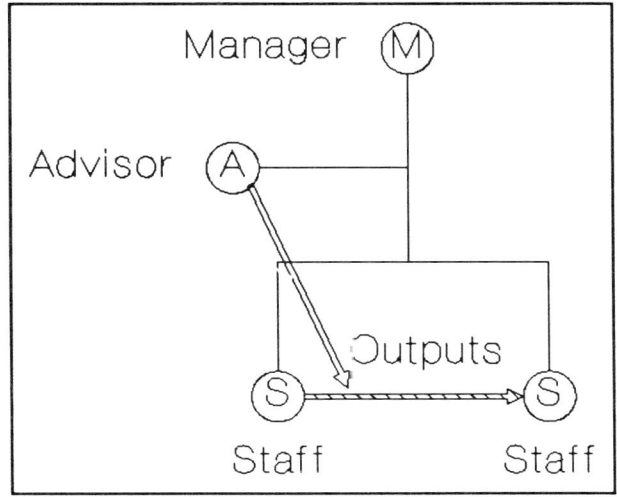

Figure 4.7 Standardisation of Work Outputs

In our view computer-integrated technologies used to pursue goals of flexibility in manufacturing have had a major impact on the balance and extent of coordinating mechanisms needed to exploit them, and these changes in coordinating mechanisms constitute the main levers for organisational change. Automation, and in particular integrated automation, incorporates much of the coordination of what were previously differentiated tasks within the structure of the information system itself.

In Zubov's (1985) terms computer-integrated technologies have two main effects: they both automate and informate. Automating has the effect of incorporating the coordination of various tasks within the information system itself, whereas informating enables people to have more information upon which to base their coordination.

We believe that the organisational changes which fit with integrated technology applied to achieving greater flexibility can best be understood through an examination of the effects on coordination mechanisms. The primary result of applying computers to manufacturing companies is to automate work which was previously done by people. A CNC machine carries out machining under the control of a program which previously required an operator to control it. An FMS extends this automation to the way the work is transported and presented to the machine, and to the way tools are deployed. A simple CAD station frees the draughtsman/designer from much of the basic work involved in, for example, dimensioning drawings. With more integrated applications of CAD/CAM automation extends into the links with production engineering and then onto the machining centres themselves. Similarly MRP models replace people-driven paperwork systems with computerisation. As CAPM systems become more integrated they enable more and more of the production management tasks to be automated. Much of the work automated in these systems is work that was previously coordinated either by standardisation of work processes or standardisation of outputs or by direct supervision.

Automation not only automates the physical or intellectual tasks, it also automates much of its coordination. Moreover, automation also extends the scope and range of tight-coupled coordination over a more disparate set of activities. By sharing common databases, establishing and utilising communication networks, and automating work processes, coordination is achieved directly through formal information systems rather than being mediated by coordination mechanisms in the social system. This constitutes a shift in formal/bureaucratic

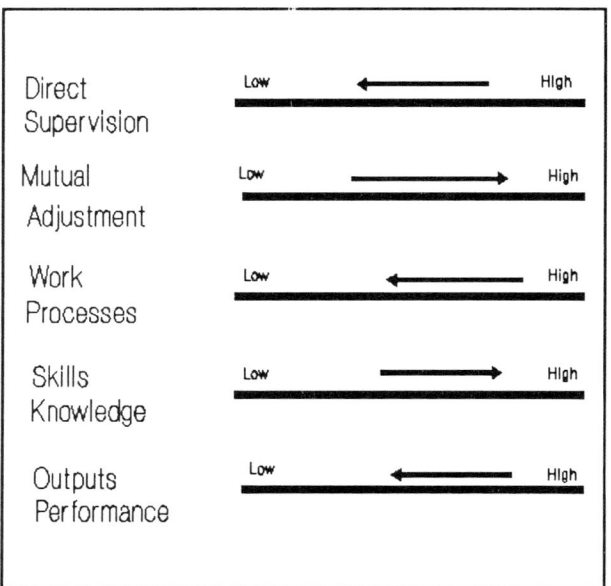

Figure 4.8 Shifts in patterns of Coordination

coordination from the social to the technical system. Some of the best examples of this in practice can be seen in CAPM systems of all varieties where coordination across functional boundaries is managed by the system in question. The overall effect on company organisation is to reduce the need for direct supervision, standardisation of work processes, and standardisation of outputs, to change the standardisation of skills from specific to integrated, and to increase the need for coordination through mutual adjustment. In Zubov's terms mutual adjustment of physically separate individuals is made possible by 'informating'.

A further feature of automation is the incorporation of hierarchical control within the system, thus reducing the need for coordination through direct supervision. This, together with decreased manning levels has the direct effect of flattening management hierarchies, and facilitating the introduction of autonomous group working. Such groups also have the advantage of using mutual adjustment as their major coordination mechanism which fits the changed role of operators in responding to exceptions, rather than continuously monitoring work processes.

Using Thompson's (1967) framework for classifying technologies, computer-integrated technology constitutes a move towards,

> 'intensive technology' which creates 'reciprocal interdependencies requiring
> mutual adjustment for their coordination'.

Thompson goes on to argue that the primary organisational design parameter in these circumstances is location, i.e. co-locating face to face workers so that they are able to adjust mutually. The increased use of cellular manufacturing and strategic business units (SBUs) is evidence of this in practice.

A New Organisational Form?

Overall, we would argue that we are seeing a change in organisational configuration, what Mintzberg calls 'superstructure'. In his terms the shift currently taking place in manufacturing is from the Machine Bureaucracy (incorporating the traditional Taylorist/Fordist approach) to the Automated Adhocracy where the administrative structure becomes decentralised and organic, the distinction between staff and line becomes blurred, communications are informal and diverse, authority is much less significant, primary grouping is by product or market, and the key coordinating mechanism is mutual adjustment. However, we do not believe that this is the whole story, for mutual adjustment operates best in face-to-face situations, yet integrated technologies have the capacity to coordinate outside of these situations.

In fact, integrated technologies through their tightly coupled systems offer the potential to coordinate directly activities which were previously disparate. To be fully exploited they need accompanying social systems which are capable of similar coordination. However, coordination through mutual adjustment in small organisational units is insufficient by itself to coordinate across functional and other organisational boundaries. It is not surprising that much difficulty is experienced in implementing integrated coordinating technologies such as Manufacturing Resources Planning (MRP II) which span the range of manufacturing activities and transcend existing functions. In these cases, face-to-face contact between those who need to coordinate their activities is limited and therefore mutual adjustment difficult.

We believe that Mintzberg's sixth coordination mechanism - standardisation of ideology - provides the means by which this can be achieved. Standardisation of ideology is where work is coordinated by people sharing a common set of beliefs about the company mission, and what is important in achieving it. They are able to coordinate their actions because of a shared ideology which defines key directions and key values. This is often referred to within companies as sharing a "common culture". Total quality management and Just-in-Time manufacturing provide such ideologies. They articulate coherent orientations to organising and doing work and bind actions together by a superordinate set of values (e.g. Zero Defects, Customer Requirements, Minimum Inventory, etc.) which transcend sectional interests. This form of standardisation is similar to the standardisation of skills/knowledge, particularly in the case of professionals who frequently become indoctrinated with implicit professional ideologies during their education and training.

The introduction of coordination through standardisation of ideology highlights the possibility of using standardisation of beliefs at the level of the company in order to engender commonality of purpose whilst using automation to support operations. This gives a dual set of levers for management to use in designing and coordinating company organisation. Morgan (1986) uses the hologram metaphor to describe organisations that use ideology/beliefs as a primary coordination mechanism. The purpose/beliefs of the

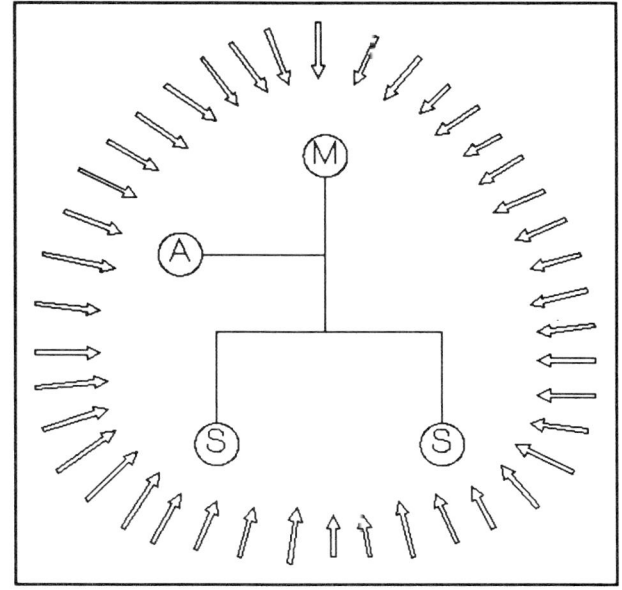

Figure 4.9 Standardisation of Ideology/Norms

whole organisation are reproduced in every part of it. The recent interest in organisational culture in the management literature which was influenced greatly by Peters and Waterman (1982), also reflects the growing awareness of company values and beliefs as an important way of coordinating action, particularly when facing dynamic and complex environments. Its strength is what Peters and Waterman have labelled as 'simultaneous loose-tight' properties. Standardisation of beliefs/ideology is 'tight' in the sense that it constitutes centralised control at the level of everyone sharing the same purpose - but it is also 'loose' and highly decentralised because it relies on individuals interpreting how that purpose is best achieved at their localised level.

It is a key task of the management team to identify, articulate, diffuse and sustain appropriate ideologies. Effective management of company ideology enables strategic coordination of activities through a common purpose which facilitates tighter coupling of the social system to match that incorporated in integrated computer systems. Mintzberg labelled the organisational configuration that is centrally coordinated by ideological belief 'Missionary'. The combination of coordination through automation and ideology is leading to a model of manufacturing organisation which is a hybrid of Mintzberg's Automated Adhocracy and Missionary forms. Given that the main feature of coordinating through ideology is a sense of common focus, we suggest that an organisational form appropriate for exploiting new technology within automated factories might be thought of as a 'Focused Adhocracy'. This seems to capture the sense of common strategic purpose, whilst retaining operational flexibility as roles are devised, passed around or shared by teams and individuals within the company. Loose, organic and emerging structures seem to provide the best context to facilitate the whole process.

In-Company Features

Whilst the above section concentrates on the theoretical formulations underpinning manufacturing organisation in the 1990s, we have been observing also the ways in which companies have been realigning their manufacturing organisation to better exploit integrated technologies. We have found it useful to look at these 'natural experiments' at three levels: work organisation, management organisation, and inter-organisation.

Work Organisation

By work organisation we refer to four areas - skills, job design, work groups, and payment systems. If we look more closely at these in practice we can see some features of the new paradigm emerging.

Past	Emerging
Single-Skilled	Multi-skilled
High-division labour	Integrated tasks

Manufacturing Organisation

Long skill life cycle	Short skill life cycle
Skill life > employee life	Skill life < employee life
Individual work/accountability	Team work/accountability
Payment by results	Alternative payment systems
Supervisor controlled	Supervisor supported
Low work discretion	Increased flexibility/autonomy

What can be seen in the moves from Past to Emerging model, is a significant shift in the management valuation of the workforce. Instead of designing and operating a system to reduce skill levels and consequently training requirements, and instead of management being primarily concerned about control, and using piece rates as a carrot/stick to induce workers to perform, we begin to see a fundamental change. The change involves seeing the workforce as the real resource in the company, a resource under-utilised to a considerable extent. The change sees the necessity for developing the actual and potential skills of the workforce and the willingness to do so.

Features of this trend can be described as a technological push towards operator multi-skilling, increased teamwork, greater discretion and problem solving lower down the hierarchy, leading to pressures to change payment systems appropriate for the new conditions.

Within companies, and particularly where the degree of integration was moving beyond stand alone applications, we saw certain features coming to the fore:

• machine operators moving from single skill machine minding to multi-skill and problem-solving practices. This included a greater responsibility for quality, and involvement in continuous improvements.

• an increase in team approaches, within and across functional departments. This was particularly evidenced in cell developments, product groups and strategic business unit

(SBU) developments.

- an increase in operators and supervisors solving day-to-day problems. This included going directly to other departments to trace problems and seek help. In this situation the line manager was used only as a last resort.

- at the fully integrated end of the continuum we saw an example of a CIM development where product builders (operators) had both the responsibility for and the skills to assemble, test, assess quality, sign and pack the product. Full kits of parts were delivered by AGVs from an automated warehouse. In this situation the line manager operated as a resource provider, rather than in the role of 'line' or 'control', and his team thus resourcing the product builder.

- payment systems were moving away from individual incentives based on output toward payment for skills, with some trying to use a group- or company-wide bonus scheme to add an incentive. Attempts to change in these areas were probably some of the most problematic which companies were facing.

With applications related to coordination similar developments were observed, with small differences which were related to the degree of functional integration in relation to the type of coordination system adopted.

- One company adopting MRPII achieved better stock control, reduced inventory, lower costs, better scheduling, better management information database. Computer keyboard skills were the main new skills introduced, together with an attitude change where the individual gained a better sense of his/her role within the system. Ownership of the system became a key role as data integrity was a key issue. Supervisory roles changed as a consequence of operators having immediate access to the database.

- Another company, with MRPII Class A user status had moved on to develop JIT using a Kanban pull system and is now attempting to develop TQM. In this situation multi-skilling and self inspection practices had developed. Operations were streamlined with a freeing up of line management time and a move away from supervision towards ad hoc problem solving.

A review of the impact on work organisation is outlined by Liu et al (1990) who argue that operators of machinery carrying out 'regulation tasks', i.e. those standardised by work processes and outputs, may have 'no direct tasks to accomplish, but to be available if some incident occurs'. The role of the machine operator becomes multi-skilled, incorporating that of machine minder, exception handler, maintainer, planner, programmer, continuous improver, etc. In these circumstances payment systems should reflect what people know rather than what they do.

Management Organisation

Past   **Emerging**

Sharp line/staff boundary   Blurred boundaries

Pyramid authority   Simultaneous tight/loose

Vertical communication   Network communication

Multi-level hierarchy   Flat structure

Bureaucratic/mechanistic   Temporary/organic

Formal control   Holographic adjustment

Functional structures   Product/project/customer-based

Status differentiated   Single status

Rigid/non-participative   Flexible/participative

Our observations indicate that aspects of structure, departmental groupings, culture and control mechanisms are changing and these changes are encapsulated in the past and emerging models as depicted within the sphere of management organisation. The emerging model contains common features of successful in-house users of integrated technology. The drive towards these changes may originate from a revitalised organisation/manufacturing strategy linked to the business strategy, and/or from the demands of the computer-integrated technologies themselves.

This latter point is worth emphasising. A common feature observed is the way computer-integrated technologies internalise the routine processes previously undertaken by managers, staff or machine operators. For example machine operators no longer needed certain skills such as setting, tooling, cutting, etc., though they still needed the basic knowledge of the processes. Instead they built up more skills, besides computer keyboard skills, of a problem-solving variety. Managers found that much of their time which had been taken up in coordinating differentiated tasks was no longer required. The computer-integrated system incorporated much of the coordination of what were previously differentiated tasks within the structure of the information system itself, extending a common database across a range of disparate activities. What we believe we are seeing is a shift of informal/bureaucratic coordination from the social to the technical system, and consequently a change in the content, and coordinating orientations of managerial positions.

As with 'work organisation', instances we observed reflected the growth of the emerging model, with some companies having developed more features than others. Some aspects seemed more amenable to change than others, with resistances tending to be more firmly expressed within the areas of functional integration, changing patterns of managerial coordination, and in the number of levels in the organisation's hierarchy.

- In one company where CAD was used as a 2D draughting facility, integrative processes were minimal. In another when the push towards design for manufacture existed, the tendency was to see teamworking developing alongside moves towards functional integration. Variations were observed in other companies from individual attempts to cooperate across functions, to job rotation, to being physically located nearer those whose work directly impinged on each other, to working in the same office. In a company developing CAE and MRP II the technical director envisaged the day when his department would amalgamate with the manufacturing department under one director. As he pointed out,

'When 95% of quality and productivity issues or problems derive from design, then the links have to be developed further and further.'

- In a company where the coordination system adopted stemmed from a top-down organisation redesign/CIM development, and where many examples of multi-skilling and teamwork developments existed, functional integration developed to the extent that only two departments existed, one supplier-oriented, and the other output-customer-oriented.

- Several companies had only three levels in their organisational hierarchy, and couldn't envisage having any more than this number. Others having six or seven were beginning to recognise the need to reduce, though some expressed being a little nervous about the consequences.

- One company, as a result of multi-skilling, and the internalisation of routine tasks and control mechanism within the technology itself, found that charge hands were no longer necessary, and nor were the foreman above or the shop superintendent above him. For reasons other than competitive edge these posts were maintained, though they were not likely to be replaced when the current occupants retired. In a separate part of the same company, where Strategic Business Units (SBUs) were being set up, both levels had been removed.

- In another company, one result of not responding to the wider ramifications of the changes taking place was the stress experienced by various production line managers. This stress specifically related to the considerable amount of time made available for them by the technological properties of the systems and by the changes in the job skills and responsibilities of their staff. Various responses could be envisaged, such as reducing the number of managers, reducing the hierarchical levels from seven to five or six, or redesigning the roles of the managers. The stress appeared to be a result of not envisaging beforehand or not taking advantage of the knock-on effects of change once it had happened.

- In the more successful companies clear signs of shifts away from status differentials to single status were to be seen; as expressed in one restaurant for all, no reserved car parking, common conditions of service, common reward systems. In one company private offices had been removed to facilitate greater collaboration and cooperation across the whole company.

- In another company that had achieved certain results but was finding the next phase more difficult to manage, stories abounded of the way management wanted everyone to work together and then continued to maintain differential conditions of service. Separate restaurants for different staff (the Golden Grill for directors, the Silver Grill for senior managers and another for 'the rest') may seem a minor issue. It wasn't experienced as minor, but as a real and symbolic statement of how top management viewed its workforce. The message was relatively simple,

  'Contradictory messages won't take your workforce with you.'

Inter-organisational Relationships

Also we have observed changing patterns in the relationships between an organisation and its customers and suppliers. Overall, in looking at the differences between the past and emerging model, we can see a shift away from confrontation and a move to more cooperative, long-term relationships.

Manufacturing Organisation

Past	**Emerging**
Tight boundaries	Blurred boundaries
Arm's length dealing	Cooperative
Short-term	Long-term
Confrontational	Partnership/developmental
Lack of customer involvement	Customer as king

- One very successful company had built up a strong long term relationship with its major world-wide customers, so much so that the customers used a JIT system as a way of placing orders.

- One company in the UK in regular communications with a company in the USA initially introduced CAD as a way of exchanging data electronically, rather than fortnightly interchange of large rolls of design drawings.

- One company, very concerned with its quality of service, brings in its retailers to train them in the exact ways they require their goods to be sold and services to be given.

- Several companies enjoyed outlining how their customers could regularly be seen standing next to the sales personnel, the design engineer, and the manufacturing engineer, studying a CAD 3D solid modelling demonstration so that the customer's design requirements could be 100% met.

- Several companies were developing computer-integrated links with main suppliers to improve lead-time and delivery accuracy, and consequently costs.

Overall, the movement toward pull systems for manufacturing organisations is ensuring a sharp customer focus and an increasingly strong set of internal relationships between manufacturing operations, marketing and product design. Other ways in which companies are re-evaluating their external relations are with suppliers of outsourced parts, raw materials, and contracts including the use of outside consultants. Whilst percentages of outsourcing can vary greatly from one company to another, a general trend which is emerging is the movement toward the development of collaborative links between focal companies and suppliers. These relationships tend to be long-term with quality as a prime focus rather than short-term and focused on cost. Ettlie,(1988), argues that firms outsourcing more of their costs have a number of interesting characteristics, not least of which is

'a business strategy that emphasises new product introduction in a more product-innovative environment...(and)...a calculated risk approach toward modernisation'.

In other words such companies had a management with a clear strategy both in terms of direction and in terms of delivery.

CHAPTER 5

Westland Helicopters

During 1987, we spent a year looking at the history of AMT implementation in Westland Helicopters based at Yeovil. Two particular applications were explored in some detail, but we had access to accounts of other applications which have provided a useful historical overview; useful in view of the fact that our contract with Westland was during a period of revolutionary change within the company. This particular study had to take account of a complicated series of events — strategic, business, corporate, technical and scientific — which had taken place at Yeovil, within the last two decades and, more particularly, within the last decade, culminating in a radical shift within the last two years.

Alterations in strategic thinking have come about in response to turbulent change, and these different stages within the company's history have in turn altered the context within which each AMT application was acquired and then implemented over the last decade. It is the intention of this study to explore these stages and, in particular, how the differing contexts affected installation and implementation of AMT.

The Situation

Organisation

Although Westland plc is best known for helicopter manufacture, it is in fact a group of technology and aerospace-related companies. In 1986 the Group employed approximately 10,000 people, and since 1985 the companies had been decentralised into four divisions:

Division	No.of Employees*	Turnover*	Profit*
Aerospace	1,500	£47m	£4m
Technologies	2,600	£89m	£12m
Helicopters/	4,500 }	£230m	£9m
Customer Support	1,800 }		

(* approx. figures 1986/7)

In 1987, Helicopters and Helicopter/Customer Support Services were made into a single Division.

The Product

The problems of helicopter manufacture can be illustrated by Westland's most recent product, the EH101, which will take approximately 15 years to bring to the market at a cost of £450m. During this time, some of this enormous cost will be offset by monies received from:

- the MoD sharing development costs which are then offset against product sales

- the DTI

- the Italian Navy

- Agusta, an Italian aerospace firm.

The costs of development of present and future helicopters are such that all projects of this nature are inevitably collaborative. In order to remain successful it is necessary to have overlapping product life cycles of development, production and sales to smooth out potentially disastrous peaks and troughs in cash flow and resource allocation. In Westland's case, the 'black hole' in the figure below, represents the cash flow crisis resulting from a catastrophic gap in the order book due (in part) to the failure of the Westland 30 model helicopter.

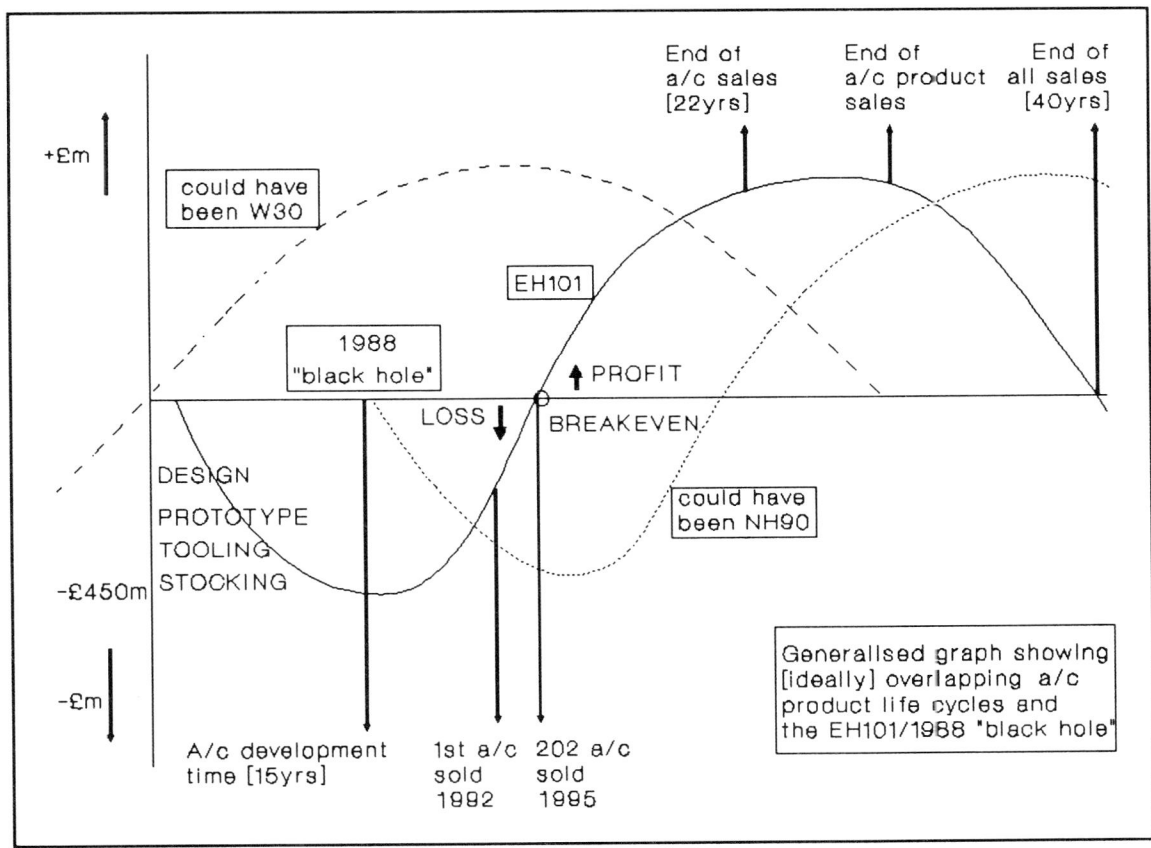

The company had invested in the development of the W30 targeted on the MoD programme AST404 and civil users. Although by mid-1984 the company had a letter of intent from an Indian customer it was unable to turn this into a firm order. In March, 1985, the MoD announced its intention not to proceed with AST404 procurement. This was said to be due to a reappraisal of airmobility strategy as a result of the experience gained in the Falklands War. This decision triggered the Westland Crisis. Having maintained full facilities to deal with the expected business, the company had high costs and borrowings and a financial crisis was precipitated by the likelihood of the October 1985 financial results making receivership probable. There was a pre-tax loss of £95m - the gearing (the ratio of borrowings to net worth) rose to 336%, which meant that the injection of new funds was essential. Following publication of these accounts and the financial reconstruction of Westland in motion, the promised £100m order from India was confirmed. The resulting reconstruction of the company brought two new industrial partners, Sikorsky and Fiat, who between them owned 14% of the voting stock.

The failure of the W30 was particularly difficult for the company since originally it was planned that this aircraft would be 'peaking' in terms of its product life cycle while the development costs of the EH101 were greatest.

It is this order book gap which has been largely responsible for the radical and continuing changes which began at Westland more than two years ago. How had this state of affairs come about?

History and Background: Developments up to 1985

In order to attempt to answer this question, it is necessary to explore company history, linking in descriptions of AMT acquisition and implementation as they occurred throughout the various stages. We have conceptualised the developments under the following headings:

- The Military Market

- The Civil Market

- Attempts to Cope with Change. e.g. the R&T Division

- AMT in the Technological Era

- 1985-87: Crisis and Developments

- 'Out of Engineering and Into Business'

- AMT in the Business Era: FMS Cells in Machining

The Military Market: Cost Plus to Fixed Price

Westland Helicopters has been building helicopters for the armed forces since 1940, and although attempts have been made to enter the civil market, it was the military market which dominated and continues to do so. The fluctuations within this market have obviously had profound effects on the company. The M.o.D moved, in some areas, from cost plus contracts ('spend whatever it takes') to fixed price contracts in the mid-1970s, with many of its prime contractors. However, developmental and repair work does not lend itself to fixed cost pricing; for instance it is often impossible to forecast how much a helicopter will take to repair until the operation is well under way. Therefore, development and repair work is still on a cost plus basis.

Since the 1940s, the company was perceived as having developed an approach to business and management which was cushioned from some of the harsh realities of the business world by cost plus MoD contracts. With the benefit of hindsight it was seen that since World War II such was the dependence of the company on the MoD that it had become a virtual MoD clone. Its strategy was determined by the MoD in terms of products and technical developments. Much of the materials, tools, working capital, designs, specifications and investments were provided by the MoD. The business style was administrative; keeping control of the budget was paramount. Westland took 8% of any contract, and so the more money spent, the bigger Westland's slice.

During this period, the prime managerial emphasis was on the technical excellence of the product. This emphasis created great pride in the product in both management and the workforce. Although changes from cost plus to a more cost-managed approach were taking place from the mid-1970s onwards, it was reported by many in the company that Westland had very little idea of how to move from one state to the next. These external changes in the environment had a dramatic effect on Westland. In 1978, the company made a loss of £19m on a turnover of £200m.

The dilemma was compared to riding two horses at once. With 60% of the business still operating on a cost plus basis, and 40% (the profit-generating part) now on fixed price, the question was asked: 'How is it possible to build a new culture on two such disparate bases?' Nevertheless, the move away from a totally cost plus environment is seen as central to subsequent events at Westland Helicopters.

In the view of many, this difficult period was made worse because Westland's experience of estimating, planning and control was, at best, patchy, and, at worst, non-existent. The company was able to estimate on the rudimentary level of predicting man hours per job, but was less successful in the planning/control area. Several 'false starts' - as one interviewee described them - had been made to try and install a master scheduling system. These had failed,

'...because the culture doesn't sustain it. Systems which facilitate planning and control often imply a sharing of power'.

This 'sharing', it was perceived, would impact on sensitive areas. It was this which led to a situation where the company was not cost-conscious, needed business management, and had a great resistance to change. A matrix type of organisational structure was imposed in an effort to react to this new situation, but in such a way as to have little chance of success. Westland ended up with seven tiers of senior management.

The Civil Market: The Beginning of the Black Hole

A strategy was developed to enter the civil market, and in the late 1970s this looked very promising. However, it is a sign of Westland's history that Bells, Hughes, Boeing and other helicopter manufacturers had been selling successfully to the civilian market for more than 20 years, whereas Westland was seen as having only an MoD strategy. Westland was probably persuaded to enter the field, by prospective sales for offshore oil exploration and commuter helicopter travel but, with hindsight, Westland did not have a suitable helicopter for this market. The W30, developed for this purpose, was an adapted version of the Naval helicopter, the Lynx, but was described by one outside critic as,

'...inferior to anything else available (in the civilian market) for ten years'.

By the time the W30 was ready in 1982, the markets had all but disappeared.

'Westland had turned up at the party, only to find that all the guests had gone.'

Nevertheless, Westland predicted a 10% market share for itself for the coming decade.

That this prediction did not materialise was due not only to the fact that the W30 was not the most desirable model on the market but also to a sales and marketing performance which has been described as 'amateur'. A former Marketing Director has been reported as saying:

'...... Westland never learned to sell .... there was only talk of mechanics .... There was a total lack of anything that makes the customer feel the salesman is trying to understand his problem ...... (there was) no fundamental understanding of transport economics, they didn't understand what customers do with their helicopters.' (*Management Today*, March 1986)

In the mid-80s, the W30 was written off at a cost of £100m. The 'black hole' was inevitably opening up, and there was no alternative product to plug the gap. At that time, the civilian helicopter market was virtually abandoned and the company began pinning most of its hopes on the EH101. However, other helicopters were on the drawing board: the NH90 (in collaboration with MBB, Aerospatiale, Fokker and Casa) and a new light attack helicopter (in collaboration with Fokker, Casa and Agusta).

It is the view of observers from both inside and outside the company that throughout this period there was a perceived unresponsiveness to the market-place, and an inability to capitalise even on opportunities which presented themselves and which did not have to be sought. A few examples were quoted to us:

- Failure to exploit the North Sea exploration market, which began to open up in the mid-1960s.

- Failure to develop possible replacement helicopters for Bristow when their Westland Wessex fleet (a substantial order in the 1960s) was grounded in 1982.

- Long lead times on design and manufacture. Westland estimates were typically longer than those of rivals, e.g. designing and installation of aircraft windows to comply with new safety regulations.

- Failure to take up an offer in the 1970s to adapt the Sea King for North Sea use at a reasonable cost, and in reasonable time. By the end of the 1970s, '... the market was saturated with (unadapted) Sea Kings, and there were 50 Sikorskys buzzing back and forth to the rigs'.

Attempts to Cope with Change: The Culmination of the Technological Era

The term 'Technological Era' refers here to that period since World War II, and particularly to events during the last two decades. However, one interviewee made it clear that the company had a purely technical culture since its formation in the earlier part of the century, which had been preserved intact throughout its various transformations.

The two helicopters which Westland had produced and developed itself (the Lynx and the W30) showed up the lack of integrated manufacturing within the business. Manufacturing requirements tended not to be anticipated. The workforce was inflexible, and, because there was a wage freeze, the workers were on piecework which was increasing labour costs. There was a firm order for 93 helicopters from the MoD and total orders of 160 in the mid-70s, but by 1978, only 30 machines had been delivered. Poor manufacturing performance was perceived to have contributed to this state of affairs, and the company was unable to predict further production schedules. An outside observer remarked that Westland was

'....over-confident in its technical prowess, it was production-led without in fact being efficient at production.'

As the company began the long descent into the crisis period of the mid-1980s, other initiatives were taken.

There was an attempt to address the issue of integration. It was hoped that this would bring to an end the 'baton passing' mentality then prevailing between different departments, and that it would also take some of the pressure off Manufacturing, which had never been regarded as a priority area. Designers were 'farmed out' to Manufacturing - and vice versa, in an attempt to reduce the design-to-manufacture lead-time and in order to foster cooperation between the departments. Previously, designers had 'taken their time' producing drawings. It was observed that this was not surprising in the context of the culture at Westland where high prestige/status attaches to Design. Because deeper issues and old antagonisms were not addressed, it was predictable that such a simplistic solution would not work. Design stalled on the production of drawings whilst Manufacturing delayed the 'can be made' signature. A cost-management initiative also failed - again because of the underlying assumptions and ways of acting of the various people involved.

A notable and typical action taken during the Technological Era was the formation of the Research and Technology Division in 1980, an initiative which seems to have been part of a corporate strategy to get the company into better shape.

The Division was seen partly as a supplier of solutions to problems which appeared in Manufacturing and Engineering, and those solutions were seen at the time as crucial to the success of the company. In retrospect, the Research and Technology Division was seen to have been the culmination of the technically-led culture which was in place at Westland. With the benefit of hindsight, the Research and Technology Division can be seen as supplying 'technical fixes', when what was needed were much more profound changes. However, R&T Division was perceived to be part of a new technological strategy.

As the R & T Division extended its experience of applying new technology, an evolution in its thinking took place. Although starting with a narrow technological perspective they began to recognise that many of the factors governing the success of implementation were organisational in nature. It also became clear as time went on that the strategic context in which AMT applications took place had a major impact on their exploitation in achieving competitive success. These lessons learned in R & T Division applied not only to the projects with which they were concerned, they had great business significance for the whole company. The importance of sorting out business and manufacturing strategy prior to AMT investment was to become a vital issue in Westland's later attempts to regenerate its competitive position.

The Division was set up to look at ways of introducing aspects of AMT into the company operations in order to improve methods of manufacture. Its formation coincided with the recognition of the growing need for the company to consider what AMT/CIM had to offer. The Division was expected to influence the managers of the various engineering and production departments, but was set apart from the design, engineering and manufacturing elements of the organisation. There was as matrix-type organisation structure, and the head of the R&T Division reported directly to the Business Development Director, who was responsible for new business.

The Division grew from 14 to 52, and seemed to have two main areas of interest, production engineering and manufacturing. They developed an integrated approach which looked at how helicopters were made and attempted to invent or buy things that would do it better, and to develop new manufacturing techniques for products specific to helicopters.

Prior to the setting up of the R&T Division, the AMT that had been implemented seems to have been restricted to NC machines. The Division, at the start of its operations, acquired seven robots (there were only 19 in the whole of the UK at that time). These were seen by the R&T Division as an attempt to demonstrate the applicability of new technology to the Manufacturing and Engineering Departments, and as a way of providing solutions to some of the technical problems.

Areas for the introduction of AMT during this period were selected on the basis of the degree of cost reduction benefits to be gained. Tabletop experiments were carried out by members of the R&T team who invited in operators and shopfloor supervisors to participate in these experiments. The following figure outlines the life of the main projects initiated by R&T Division:

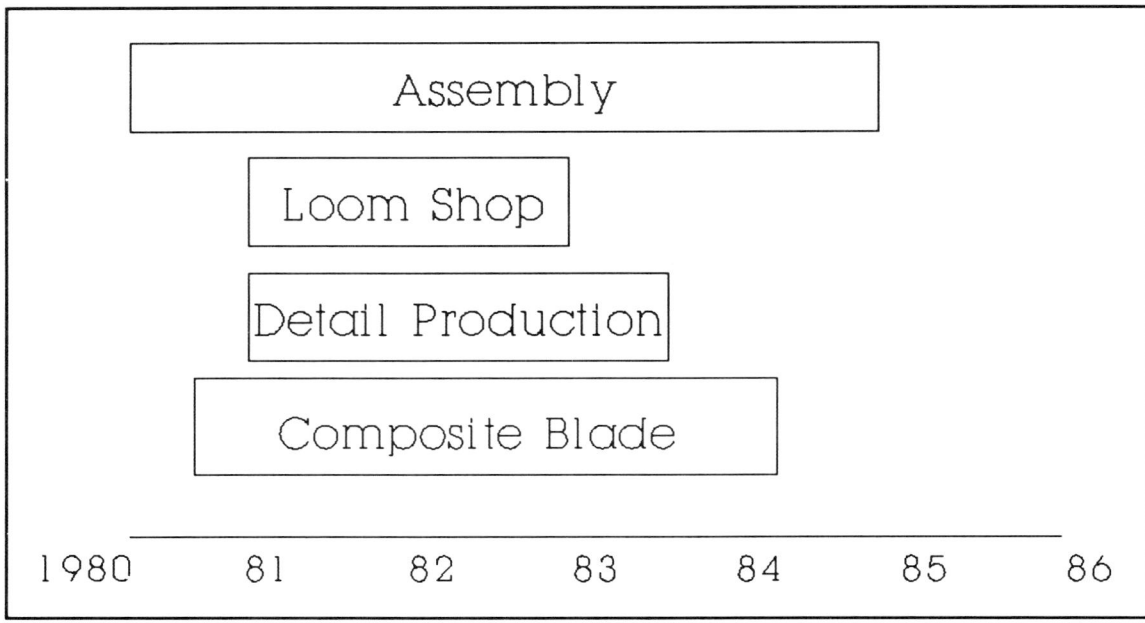

A CAD facility was also installed which allowed for dynamic and stress modelling. One opinion was that CAD by itself was not profitable. 'A CAD drawing is an expensive one; to be truly effective CAD must be linked with CAM'. It was felt that, had a more strategic approach been taken, such as a possible investment in programs which would have allowed the company to model the market, then this course of action would have been much more cost-effective. The market analyses produced would have facilitated the crucial 'make versus buy' decisions at a much earlier stage in the company's history.

AMT During the Technological Era

This section explores the various AMT installations and implementations during this period. One project, composite blades, is recounted in some detail.

Assembly

Assembly in the aerospace industry is a low-batch/high-complexity/high-variety, manipulative environment which offers little scope for introducing automation or mechanisation. It was thought that the main way to contribute to effectiveness was to ensure that the assembly task was a straightforward job, not a complicated one. It was estimated that two thirds of the assemblers' time was spent figuring out what to do, often finding they had an incomplete parts kit or that they were unable to do the job due to the prior/adjacent assembly being incomplete. In this situation there was scope for better information systems which would help do the job by providing a remote sorting of the task, schedules for full kit, and production management information.

The view was, and is, that there is no justification for automating assembly in the case of helicopters. The idea is to develop JIT techniques for assembly using manpower rather than machinery. The presentation of information is important here because of the need for flexibility, and the intention is to reduce costs by reorganising the information flows.

The Loom Shop

Installing the wiring loom of a helicopter is a complex and time-consuming task. Therefore this area was chosen; firstly because of the likely cost reduction benefits, and secondly because the work was thought to be mundane and repetitive and made up of simple tasks. Models were set up to demonstrate a 35% cost reduction. What had not been considered, however, was the part played by the operators in terms of the information handling they carried out. This threw up the question of whether loom manufacture could be mechanised successfully, and led to the R&T Division requiring the cooperation of the operators. This was not forthcoming to begin with, but, as the project progressed, a genuine collaboration began, and for the first time there was a realisation of the limitations of a technologically oriented approach to design and implementation issues. The loom shop experiment became, accidentally, a collaborative, participative project. Four shop-floor operators, one inspector and one R&T specialist worked together to put forward a mechanised loom unit. ELECTI CAD/CAM was utilised to make the loom shop paperless.

The loom project was thought to be successful in that the original expectation of 35% cost reduction was met. This was a reduction in the operating cost of the department itself. In addition, the nature of change on particular jobs and the high degree of participation were thought to be other aspects of success.

There was a design 'spin-off' associated with the loom project in that feedback was provided to the CAD system via terminals on the loom shop-floor, which has allowed the loom runs to become part of the design process. This would appear to be quite beneficial because a number of changes are made by customers to individual aircraft.

The Detail Shop

Here sheet metal parts are manufactured, also in a paperless environment. An expensive production complex was developed whereby, via CAD/CAM, the instructions generated by a drawing on an electronic drawing board are hard-linked to the manufacturing facility. The instructions are down-loaded onto the shop-floor and converted into routing/cutting procedures, leading directly to heat treatment and other operations prior to packaging and despatching of the correct parts ready for assembly.

This application of CAD/CAM in the detail production complex did give the company a 'slick' manufacturing facility, but later views came to suggest that,

'simple tin cutting of this kind could be subcontracted out.'

Initially it was installed because,

'the engineers liked it, and the directors liked the idea. Eventually, it acquired a momentum of its own.'

However, later rationalisation of the business led to a severe questioning of the justification and utilisation of this once lead-edge application.

The Composite Blade Shop

What and Why? The AMT: Its Rationale

The context within which this AMT application took place was a troubled one. As one observer remarked, although Westland Helicopters had to get into composites, (composites were being demanded by the customer) this 'getting in' had not been thought about in any way which could be called strategic. Its relative success could be regarded as partly fortuitous. In total £6.12m was spent on the composite blades facility.

The Manufacturing Research and Development Department, a forerunner of the R&T Division, was involved, along with other companies, in a Royal Aircraft Establishment programme, in basic composite research. The idea was that a dialogue based on results should ensue, and that the best way forward could be planned. The development of composite technology was very fast. For example, in the early 1970s, when the Lynx rotor blade was being designed, to have used composites in its manufacture would have been a high-risk strategy given existing knowledge. This is unacceptable in aircraft design. The decision was made to use stainless steel. Two years later, the state of knowledge in the field enabled Aerospatiale, one of Westland's competitors, to design and manufacture composite blades for their Gazelle range. By now this was a low-risk strategy.

Quickly, composite blades were being demanded by the customer, chief among whom was the MoD. In the late 1970s, an investigation into the production of composite rotor blades was jointly funded by Westland and the MoD. There was an input from Westland's Research and Technology Division and, by now, this technology was shared throughout the industry. Composites were developed which had immense advantages over metals. For example, metal fatigue life was 1000 hours, whereas that of composites was unlimited. The MoD was now insistent that the new material should be incorporated in all future products.

In changing to the new technology, the difficulties have been summarised in a paper written by one of the Westland employees to whom we talked:

- Design in composites is infinitely more complex than in metals.

- A composite blade has 10,000 items of WIP, as compared to the 150 items of its metal counterpart.

- The new composite materials were (and are) more expensive than metals, and of the order of four times more expensive. This changes the Labour:Material ratio from 80:20 (metals) to 20:80 (composites).

As detailed in the first two points, the inherent complexity of design and the control of the large number of parts meant that there was no viable alternative to automation. Set against these 'disadvantages' are the following points:

- Infinite scope is offered to the designer in terms of engineering performance.

- Composites have an infinite life compared to metals.

R & T Division demonstrated the feasibility of making composite rotor blades in a situation which was almost like 'laboratory conditions'. The translation of this 'demonstrator project' into a fully functioning production plant provided Westland with valuable learning on the conditions for successful implementation. In 1983, a new manager was appointed to set up the new production facility. From the outset he took on personal ownership of the project and the process. He personally planned the whole facility. Starting from a core process and an output requirement he set up a JIT manufacturing flow line, planned the type and size of building, the equipment needed, and the control systems. He project managed the building of the plant and the recruitment and training of staff. The new production facility was an outstanding success and represented a step function change from metal blade production techniques and methods. Members of R & T Division felt that a large amount of the success of this project could be attributed to the manager taking ownership of the implementation. (See below for more details).

The following facts and figures illustrate the step change from metal blade to composite blade production:

- Inventory turnaround eight or nine times per year (rest of factory once every 14 months)

- Minimum WIP (in certain areas)

- The manufacture time of composite blades is much quicker than that of metal blades

The process of blade making has two main stages:- tape laying and steam pressing. The raw materials for composite manufacture are supplied JIT and held for short periods in a cold store (the materials deteriorate at room temperature).

The blade shop is laid out in bays where each process of building up the blades is carried out in turn. Each bay is computer-linked to the store-room, and a precise count of materials used each shift is keyed into the terminal, then relayed to the stock room, and the correct amount of materials delivered for the next shift. This is an internal JIT 'pull' system, and obviously people are very pleased with its operation. This, as it was remarked, was a line, not a batch system. The main expenditure here had been the tape-laying machine; before automation this had been a very labour-intensive operation. The company had looked around the suppliers but in the end had designed its own machinery. The current machine is the Mark IV version.

Production control was largely designed into the JIT flow-line. However, an information system was needed to plan resource allocation and also the business performance of the shop. Software was designed in-house and was driven by a VAX computer. Within the shop the plan was that every process should be performed on every blade, on every shift. Each bay deals with a particular type of material and the counting of materials every day generates a pick list. There was strongly held opinion that manufacturing systems should be designed for the process needs in question. The final control system which was implemented was able to reveal how much money had been made per shift. This essentially in-house development was seen to have major advantages over turnkey packages which do not have the capacity to control vital idiosyncrasies.

The four steam presses through which the blades then pass are fully computerised. They had each cost £2.8m and were performing very satisfactorily, with few problems. After pressing, the blades were inspected and tested ready for dispatch to customers. The major benefit of the blade shop investment was that batch production was not thought to be suitable blade for manufacture,

'you can't have batches of ten blades lying around. High cost WIP had to be eliminated as far as possible and expensive inventory reduced'.

This philosophy was certainly working in one area and hopefully would extend to the rest of the shop when the problems referred to earlier had been ironed out. The end result would be cost reduction, higher quality and the production of the kind of blade the customer wanted, at the time it was wanted. Certainly, the attack on lead-times had been impressive. The comparison was made between the production of a blade and that of a gear.

$$
\begin{aligned}
\text{One blade} \quad &= \ 80 \text{ man-hrs} \\
&= \ 6 \text{ weeks (start to finish)} \\
&= \ 30,000 \text{ (sold for)}
\end{aligned}
$$

$$
\begin{aligned}
\text{One gear} \quad &= \ 60 \text{ man-hrs} \\
&= \ 18 \text{ months} \\
&= \ 1000
\end{aligned}
$$

The facility could be scaled up to provide 30 times the output produced at the moment.

How? Implementation and Installation

The manager of the Development project had been influenced by books he had read on the subject, and had his own clear ideas. During the early stages a core of people had been identified. These people had contributed ideas and had gradually become familiar with the processes and the equipment. This equipment was gradually introduced into the new production area which then took over the process from Development. This took five or six months.

Every time something new was introduced from Development to Production a core man would train three or four others alongside him. Many of the improvements came from suggestions provided by the men themselves.

The Development Project Manager had to justify the expenditure, raise the money, talk with architects, plan all the equipment details, decide manning levels, and explain to the men who were going to be involved.

These explanations were seen as vital to the successful implementation of the new blade shop, over and above the technical training provided, and were carried out largely on a one-to-one basis - especially with the supervisors 'coaxing, talking ...'. There could be no doubting the enthusiasm of the manager, and the fact that he could probably be very persuasive because of his belief in the cause. There was a positive perception of the facility and the way it had been implemented by both management and workforce. Clearly the implementation had been handled well, but also there was a general realisation that the advantages produced by a streamlined facility were in everyone's interest. The message 'once per shift' had been a valuable one in getting different working methods instituted. Items of news such as the fact that composite blades produced at Westland might actually be used on the U.S. President's helicopter were a,

'tremendous motivator and lifted morale'.

In response to the question: 'What would you have done differently', several points were raised:

- Attention must be paid to detail. There were so many interdependencies within the system. You can get all the big bits right and then find out that small items foul up the whole system.

- You must not forget about the people. If you just concentrate on the AMT, the machines will not produce. In the manager's opinion many FMSs had failed for this very reason. People's jobs would inevitably be changed, but those affected must be involved in that change.

- Do it, try it, fix it; don't analyse the thing to death. Once you're decided, do it quickly.

- Don't be afraid of mixing high and low tech.

- Don't necessarily use textbook solutions.

- Don't use something just because it's in vogue.

Developments in the future might include taking a stronger line with the suppliers. In 1987 Westland employed 200 in-house quality inspectors, but only six such inspectors were sent out to check on the quality of their suppliers. Sixty percent of the helicopter costs are bought in and it was felt that they were a large enough company to be able to ask their suppliers for 100% quality. These quality expectations would in turn be matched by Westland's provision of 100% quality for their own customers. It was observed that the concessions which were being raised in the blade shop were not indicative of 100% quality. 'We need Total Quality Management' was one remark.

It was also perceived that in the management of their customers several aspects could be improved. The customer should be asked what was wanted, and what small changes needed to be made before delivery. The customer should also be told when Westland had instituted some practice directed towards an improvement in customer relations.

1985-87: Crisis and Developments

The account which follows incorporates not only views from within the company, but also descriptions and opinions given in newspapers, journals and magazines. Although Sir John Cuckney, appointed as Chairman of Westland plc in 1985, has referred to the press in his 1986 report as providing for Westland,

'... far too much publicity, much of which was thoroughly unhelpful to the company's international standing.'

This amount of publicity could be said to be a mark of how significant and important Westland Helicopters is felt to be.

Crisis

When Sir John Cuckney was appointed in June 1985 he found that Westland, although renowned for making helicopters, was unable to sell them. The company owed the banks £80m, some of which debt was as a result of building £42m worth of W30s in expectation of an Indian order that materialised only years later. Cuckney reported that the company was virtually bankrupt, and until Christmas of that year he was involved in a 'race against time' to produce a rescue package mainly with Sikorsky, but bringing in Fiat 'to give the deal a European flavour', as reported in the press. He was successful in bringing off this deal in competition with a rival all-European bid (British Aerospace and GEC, UK; MBB, W.Germany; Aerospatiale, France; Agusta, Italy).

The Westland Affair

The debate and eventual row as to the respective merits of these bids brought about the resignation of two Cabinet ministers, and caused doubt about the survival of the Government itself. Sir John Cuckney saw the deal as being a 'private sector solution' but, given the heavy reliance still on the military market, doubts have been expressed about this. The affair was seen at the time as centring on whether the UK's independent helicopter maker should fall under US control. The Sikorsky/Fiat package involved an injection of £75m and a promise of two million man-hours of work from Sikorsky. Cuckney apparently saw Sikorsky's Black Hawk as partially filling the 1988 black hole, since it was hoped that work on some of the Hawks (e.g. ten rescue helicopters for the Swedish Defence Ministry, and other orders) would be carried out at Yeovil.

Decentralisation

The new Chairman identified that Westland owned a collection of aerospace and aerospace technology-related companies which were making money. He decided to build up a new Westland based on splitting the company into divisions, and the company would be taken along a route dedicated to aerospace as a whole, rather than exclusively down the helicopter track. On a £230m turnover in 1986, the helicopter and customer services divisions made a joint profit of £9m - which, compared with the performance of the other divisions, is tiny. The Aerospace Division made a profit of £4m on a turnover of £46m, and the Technologies Division made a £12m profit on a turnover of £89m.

In Spring 1986, Cuckney organised the company into the four divisions mentioned at the beginning of the report. Of these divisions, Helicopters was the most problematic. The market is uncertain, and Westland is in an exposed position. Its major European rivals in helicopter manufacture are all subsidiaries of national aerospace interest, firmly lodged in the public sector, where profits are not the top priority. MoD requirements have not been clearly stated in the past, and as the new chairman himself said,

'Coherent planning against such widely moving (MoD) target dates, is frankly impossible.'

Its US competitors do have to make money, but they are all parts of giant conglomerates that can afford to make long-term decisions on products or marketing. Westland enjoys no such luxury. It is perceived as a strong engineer with a poor sales capacity. The urgency to complete the transition to strong profitability is pressing.

By the end of 1986, profits for the group as a whole stood at £26.4m as compared to a loss of £95.3m for the previous year, and it was felt in the company that an encouraging improvement had been made. In January 1987, a new Managing Director was brought in, and a reinvigoration of Westland's marketing strategy was initiated - advised by Sikorsky.

The Rescue Package

By early 1987, there were still not enough orders to fill the alarming and looming order gap, and Westland were awaiting the results of the publication of the UK Ministry of Defence Joint Chiefs of Staff report on helicopter requirement. There were still no orders for the Sikorsky Black Hawk which Yeovil hoped to assemble.

The MoD orders, when they were revealed, amounted to £100m less than expected, and the £300m 'rescue' package was seen as 'minimalist'. Although the requirements contained an order for 50 naval versions of the EH101 which was larger than some had expected (38 had been identified as needed), the other orders, for 25 utility versions of the EH101, 16 Lynx and seven Sea King helicopters, would be insufficient to plug the order gap and would provide only six months' work.

The package also contained the news of the Government's withdrawal from the five-nation NH90 which was being developed for the mid 1990s and was seen as an ideal replacement for the Sikorsky Black Hawk and French-built Puma at the turn of the century. This was seen as an inevitable outcome since the Sikorsky deal and further evidence (according to some views) that the original American rescue deal had failed. The official Government version was that the NH90 was 'no longer needed', but inside sources, reported in the press, suggested that there were 'strong financial reasons' for the withdrawal. press reports the following day can perhaps be summarised,

> 'Westland seems to be the victim of an inconsistent and unpredictable MoD purchasing policy, but some of the blame for the 'order gap' must be put at management's door. Perhaps in the past Westland had not given the Armed Services the kind of helicopters it needed. However, even by this minimalist 'rescue' package, the Government has tacitly accepted the case for a largely British-controlled helicopter manufacturer.'

Further Reorganisation

The strategy now was reported as cutting the size of the helicopter operation to fit the size of the market. 2000 job losses were announced in order to remove £20m in overhead costs. Many of these losses were at the Weston-Super-Mare facility, whose Helicopter Customer Support Division was to be brought over to Yeovil. Managerial reorganisation was also taking place.

'Out of Engineering and into Business'

The title of this section is a quote from one of the managers with whom we talked, who was describing the much tighter structure and management style which had been produced since Sir John Cuckney's installation. This in turn has resulted in a much more business-oriented and partially profitable business. The pre-tax profits for the half year to 31 March 1987 had been pushed up to £16.2m compared with a previous loss, and deliveries in the half year were sharply higher at 18 aircraft, compared with just five in the same period the previous year.

The Business Strategy

A positive approach had been taken towards this drastic restructuring, as reported by several of the interviewees. The company has now produced a five year business plan. The senior executives have produced an equation:

$$\text{COST} = \text{Selling Price} - \text{Required Profit}$$

If, after examination, this equation does not work out with reference to a particular project, then that project is not proceeded with, no matter how attractive it might be technologically. Part of planning for the future business includes targeting production costs, designing to those costs, and reaching the crucial 'make versus buy' decision. If 'make', then what are the resources and activities implied? If 'buy', then what are the required specifications and price?

In order to be able to make the correct decision, as defined in the preceding paragraph, the following chain had been identified:

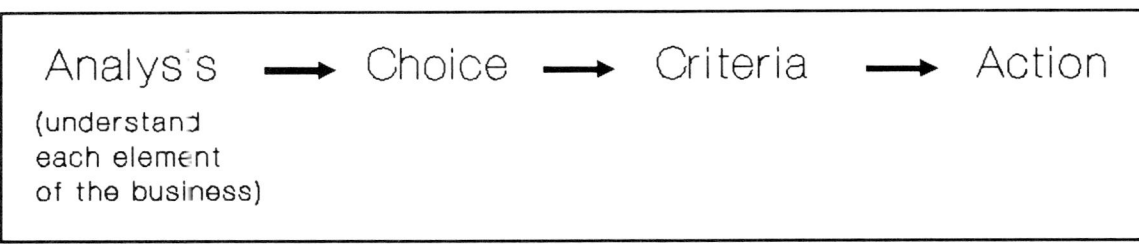

Analysis → Choice → Criteria → Action
(understand each element of the business)

The severe workforce reduction was perceived as a problem by some within the company. There might be a loss of 'critical mass' - the power to perform those tasks considered crucial to the business.

Tasks Crucial to the Business

It was inevitable, and widely recognised within the company, that the new technology would lead Westland Helicopters in the direction of producing whole battle systems - not 'just' helicopters.

In order to remain a viable helicopter manufacturer, the company has stated that it needs to retain control of:

- systems integration
- the design process
- the manufacture of all key items (e.g. gears and blades)
- the assembly of all key elements
- the assembly processes.

Key items were described as the 'sovereign bits', and because of the severe manning reductions, the company would have to make sure that it still retained the right men to deal with these parts. For a variety of reasons, such as squeezing on manpower, lack of resources, the retention of the 'sovereign bits' would inevitably push the low technology area out of the company.

In the battle systems area, EH101 was referred to as 'a flying command post', the design of which had involved the people concerned in quite new technological fields. This expertise was seen as crucial to the company's stance in the future, and in order to enhance their performance in this area, a small software house had been jointly acquired two years ago by Westland and another firm. The software experts had been working alongside the staff at Westland Helicopters, and the acquisition was seen as contributing greatly to their expertise in this area.

It was thought that the skills required by prime contractors for the MoD in the future would lie with the software houses, and it was perceived that in this area, as far as systems went, Westland would have to be able to compete with big defence electronic companies such as GEC, Ferranti and Racal. The history of the company's relations with the MoD had been one of informal networks, but now the new people in the Ministry understood the technology which would be required of a prime contractor. Once nominated by the MoD, such a contractor would be able to control and buy what was needed,

'.... from everyone else. We don't want to be "everyone else"'.

AMT In the Business Era: FMS Cells in Machining

As with the composite blades shop, it is the intention here to concentrate in some detail on the installation and implementation of a particular AMT application during this later period.

What & Why? The AMT: Its Rationale

This AMT is beginning to come on-stream at the time of writing, in the context of the 'tighter structure and management style' referred to earlier, which is producing a much more strongly business-oriented company. This orientation is reflected in the way in which the FMS has been planned and linked into the business plan.

100

During the planning phase there was a 'gut feel' that of the two main types of solution needed for machining prismatic parts - a pallet pull or a more integrated FMS - an FMS would be preferred. With pallet pull six or seven machines would be needed with high utilisation, and there would be a need for a large number of cutters and a high degree of commonality. This would be very expensive. Each of these machines would need manning. It would be cheaper to go for an integrated FMS.

A decision had been taken to invest in a number of FMS machining cells. At the time the case study was carried out, capital approval had been gained for the three cells and they were in the initial stages of installation. The manager of this area was convinced of the need for a manufacturing strategy linking business strategy to operational manufacturing decisions. Given the company's cash flow problems over the next four years, it had been forced to reduce manning. With a reduced manufacturing capability, the decisions on what to make and what to buy had become even more critical. The decision to keep control over the 'sovereign bits' such as blades and gear boxes clearly dictated part of the manufacturing strategy, but other decisions needed to be taken on straightforward cost grounds - e.g. buying in those parts that could be made elsewhere more cheaply than at Westland. It was also thought that investment in new technology was part of the price of staying in the 'hi-tech' market; that one did not look a credible manufacturer in the aerospace business in the eyes of potential customers unless one was investing in the most advanced, 'exotic' equipment.

An analysis was carried out of the existing machining facility. It revealed that there were 19,000 part numbers on the shop-floor. A parts analysis on added value showed that 80% of part numbers incurred 25% of costs, and 20% of part numbers, 75% of costs. The analysis also showed that of the 19,000 only 3,000 were 'active' part numbers. Of these 3000 active part numbers it was decided to make 400 and procure the rest outside. Similarly it was decided to procure the inactive parts when required.

A situation had arisen in the machine shops where there were a lot of older machines which were there to machine parts that were only required very infrequently. This resulted in a high level of machining redundancy with utilisation levels as low as 10%. By deciding to procure infrequent parts and make only 400 of the 3000 active parts it was possible to get rid of almost all of the old machine shops and replace them by four FMS cells. These FMS cells were capable of producing low lot sizes at no cost penalty which suited the low production runs (sometimes only five of a certain aircraft were produced). The basic configuration of the four FMS cells was determined by an analysis of the 400 parts and it was decided to 'shop around' suppliers to get the best fit machines to meet the required specifications.

'We went around the tool people - some of them didn't understand our questions, let alone give us answers'.

It was felt too that the company was not too sure of the questions either, and that it and its suppliers were involved in a relationship of trust at the end of the day, exploring the situation together. Three FMS machine cells with extensive packages to deal with the prismatics are coming on-stream, to be followed eventually by a fourth. The first machine, costing £3m, was already on-stream, at the time of writing. The eventual expenditure would be £10-15m.

How? Implementation and Installation

As the FMS cells came on-stream, there would be a need to reduce the labour force from 400 to 150 over a period of two to three years. Notes with full details were delivered to the workforce. There were slides, talks, films and discussions about what was being attempted, and where the company, or at least the machine shop was heading. It was explained that unless operations were rationalised, then an edge would be lost, and there would be lack of investment. It was also explained that there could be further redundancy, but that the idea was to try and use this rationalisation and new technology investment as an opportunity and a firm base for future growth.

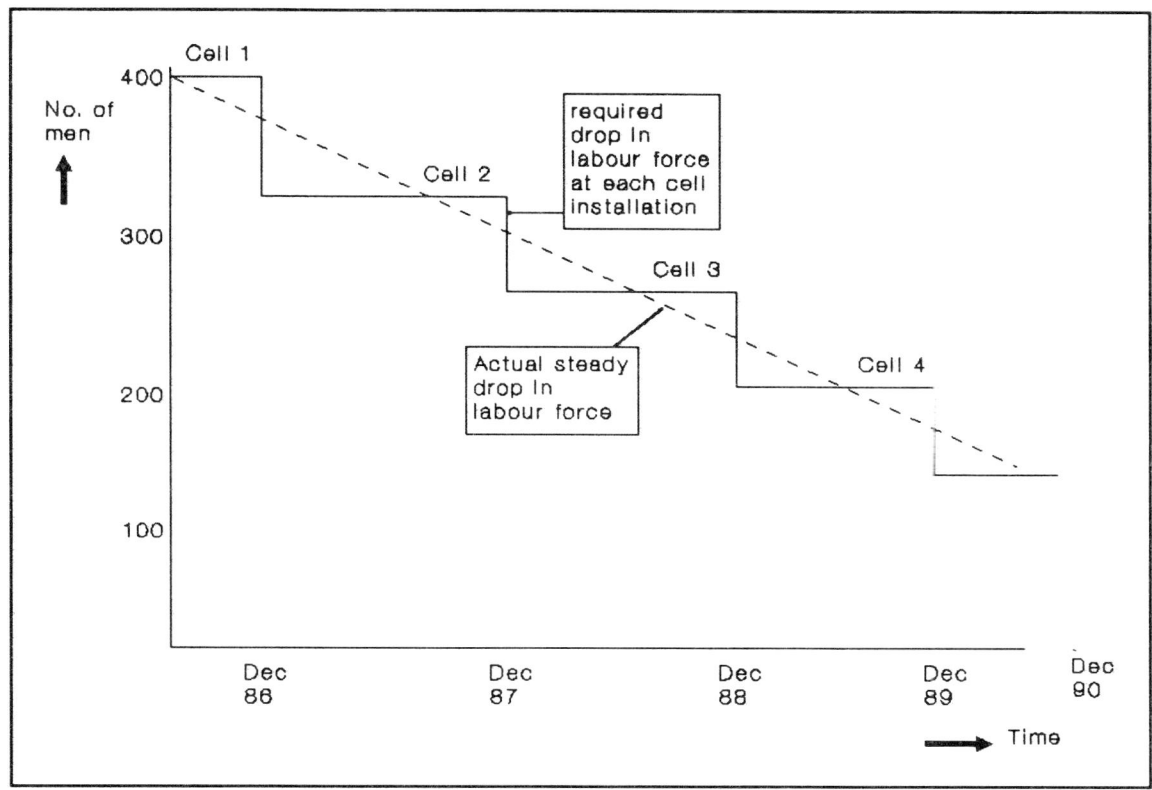

A few months later there were fuller, clearer explanations, brochures and details were handed out, and trade union cooperation was invited. Management was giving a lead, but many shopfloor ideas were incorporated into the plan.

The policy was implemented by redirecting manpower, e.g. non-essential skills, apprentices, etc. It was decided that a steady decline in workforce numbers - about three a month - would be much more acceptable than losing a great lump of 60 every time a new cell was installed. This policy was followed, even though it meant that from time to time, the shop was actually short of operatives.

When questioned about attitudinal changes which might be required in such 'step changes', one interviewee remarked that,

    'it seems a terrible thing to say - but buy the equipment first'.

This, in a way, forced through change. However, it seemed as the interview progressed, that integrative thinking had been going on well before technical installation. Important people in this whole process had been recognised early on as the production engineers.

'production engineers are the actual key; we have treated them like battery hens in the past. Here, we need their creativity, we have to say "we need something like this - go away and do it".'

Prior to the decision to 'go FMS', a small group of production engineers had been targeted, and their ideas concerning the manufacturing strategy, and the best solutions to underpin that strategy, had been solicited. This group had been identified by the Operations Manager as one in which creativity and forward thinking were strong points. The official project team comprised:

- Chief Plant Engineer
- Shop Superintendent
- Jig and Tool Draughtsman
- A Production Engineer

It was reported that early attempts at decision-making were not very successful. Regular meetings were held, but little progress was made. At this point, a smaller committee was brought together whose task it was to examine alternative solutions, particularly in the important area of altering control systems, and the need for what were described as new ways of working. Again, the start-up period of this smaller group was reported as having been

'... terrible. Someone would put forward a plan, only to have it shot down in flames by the others. For two months, there seemed to be no progress at all.'

During this period, short (ten minute) meetings were held on an 'ad hoc' basis. In the third month, it was felt that matters began to improve. This improvement was fostered by a certain sense of urgency arising from the realisation that the new technology would be brought in. Within the workforce in general, various incidents served to reinforce the new ways of working. A particular example was given: a good operator had been presented with a newly-designed job plan by the Chief Methods Engineer. The operator was of the opinion, it was reported, that the plan was no good and that the job could not be done in that way. The CME had virtually had to say, 'you will do it'.

The job plan proved to be successful, an improvement on previous ones. This produced a complete change around in attitude. The perception was that from thenceforward,

'... people began to see the light.'

One interviewee remarked that it was very important for management not only to provide this kind of direct lead, but also to be capable of being influenced and much more open than in the past, fostering innovation and encouraging feedback.

Financial Justifications

The MoD, which is still Westland's major customer, requires its suppliers to base costs on what is known as the 'composite hourly rate'. The composite hourly rate for a product is the total cost of making the product divided by the total man-hours involved. The major problem with this measure is that it does not distinguish between low and high capital-intensive manufacturing, nor does it take into account variations in the efficiency of inventory management. This has the effect of making the justification of AMT difficult and provides little incentive to adopt JIT methodologies. The composite hourly rate measure has a particular impact on manufacturing strategy make/buy policies. In the view of more than one

105

interviewee, these methods were outmoded and produced distortions. The firm control which the MoD now kept over its budgets was seen as due, in part, to its having been 'taken for a ride' by some of its prime contractors in the past.

However, it was stated that since the MoD was a major customer, Westland must make the attempt to manoeuvre within these constraints and persuade the MoD that with AMT, for instance, inventory can be reduced, especially in a business such as Westland. The following example was given as an illustration of this:

'If things were costed out wholly in the traditional way, the company might never invest in any new technology. A case had to be made to the MoD which persuaded them that with AMT, inventory is reduced. The inventory and equipment involved in assembling a helicopter are extremely costly in themselves, and extremely costly when left lying about or not being used. The more quickly high value-added parts could be manufactured and progressed through the factory, the more quickly the company could begin to reap the profits. With some of the present technology in place, and the methods of work which were dictated by this technology, something like the Lynx helicopter which the company sold for £4m could take 30 weeks to assemble. During that time, the equivalent of £30m in parts and machinery were 'gathered round it'. If times could be shortened, if methods of work could be altered (e.g. JIT reducing WIP and inventory) then "........ savings to the company would be enormous".'

AMT could provide that capability.

Other complications impact on the company because of its special position vis-a-vis the government and Defence, and because of the complicated collaborative arrangements which are a fact of life in this area. There were no 'simple' decisions in their financial world. Even the question: 'Is it cheaper to make it here than outside?' is complicated by the fact that there are 'off-set' deals with other companies, or the government has 'off-set' deals with other countries. Simple decisions cannot be set against a known base.

Within the company itself, the financial people often seemed to 'miss the point'. Every detail is examined very critically, but, as one view had it, their justifications had, in the main, been passed by the Finance Department, and people involved in making those justifications must avoid getting trapped in 'financial trees'and making their own plans so detailed that they could not see 'the wood'. Standing back was essential. One interviewee said that Westland had to get its own broad, systemwide plan and then break it down into justifiable details, not the other way round.

Amortising set-up times with run times was a problem. This inevitably led to large batch sizes which in an industry such as Westland, with small runs of customerised products, carried their own problems. The questions about attempts to reduce set-up times was not pursued by the research team.

In the lead technology area, the control of procurement was of central importance. As already referred to, 70% of the whole aircraft is bought out, and it could be said that Westland has plenty of experience of this. However, there had been a question of quality in this area. Westland exists by customerisation, and deals in very small batches. The interface between a company and its suppliers can be difficult to manage, particularly so between a process (supplier) industry and a small batch (customer - Westland) industry, and this is seen by some as being a difficult procurement situation. The company could be 'squeezed' (i.e. economies of scale might be denied them). There were still some elements of cost plus mentality around where the cost of bought-in items needed in the hi-tech area into which the company was moving would be very high. A close eye had to be kept on this. It was felt that quality of these items was absolutely essential, and that the company must be able to rely on its suppliers.

Learning and Generalisations from this Case

This case study has encapsulated many of the problems which companies typically find when installing and implementing AMT, and well illustrates the close connection between the way an implementation is carried out and the context/climate prevailing within an organisation at any given time. Although only two examples have been described in depth in the case study, and these are relatively recent applications, the researchers have had access to descriptions of other AMT applications, which enable comparisons to be made with respect to applications in their varying contexts.

Clearly the context in which various applications were implemented over the period reported in the case has had considerable impact upon the success of the process. The technological era produced both strengths and weaknesses. Characterised by the strong engineering emphasis, using R&T Division as a specialised non-line function and demonstrator projects as its main method, this era showed the strengths of the company in technological development and installation. Indeed, once the technology was justified, the R&T Division had shown that technology is relatively easily installed and that technical problems are largely resolvable.

During this time when Westland Helicopter Division's view of itself was that of a specialised engineering firm making whole helicopters, the culture of the firm emphasised technological pride. The firm saw itself as making the best helicopters in the world. Therefore it is no surprise to find that the AMT applications at that time were technically-led. Anything which improved the ability of Westland to make whole, and ever more advanced models, was, by definition, a good thing.

The Achilles' heel of this era was the lack of importance given to how the technology impacted upon business performance, and how it was to be integrated into an organisational form which could exploit it to the company's best advantage. The organisational implications of AMT were later to be seen as critical for successful implementation, and nowhere was this better demonstrated than in the loom shop project.

Other applications were less successful. Examples were provided of what could, in retrospect, be regarded as implementation failures and/or outward signs of the culture.

- As has already been described, the applicability of the CAD facility has been questioned, and it is also reported that consideration is being given to selling off the application in the detail shop, since 'tin cutting' could be bought out.

- At a time when there were only 19 robots in the UK, seven of them were in Westland. These were never intended to be used in Manufacturing - but as evaluators of technical, manipulative problems for the R&T Division.

- The story of the riveting machine introduced into the sheet metal section is also revealing and is reported here since it provides its own commentary. On a trip to the US, a machine — capable of riveting ten times faster than a man — caught the eye of two people from the R&T Division. '"We'll have one of those" we said - it cost £40,000.' The plan was that, in order to utilise the machine effectively/efficiently, structures (for riveting) should henceforth be designed for automation, i.e. designed around the new machine. 'One guy played around with the machine for a year then wrote a book on it, a manual. It was an excellent book, all about how to use it and so on and so forth. The riveter was unveiled with a great fanfare and flourish to Manufacturing and Design. They just didn't want to know.' As the story-teller admitted, 'We should have known better.' The user departments had at no time been involved in the acquisition, the planning or the installation. The machine was regarded as a threat and stood idle for a number of years until an engineering enthusiast, one of the user groups took it on, and the necessary ownership was forged. Forty per cent of the riveting in this area is now carried out on this machine.

These implementation failures, coupled with the impending business crisis and change of government in 1979 reinforcing trends away from 'cost plus' approaches, could be seen as contributing to change in the ways in which AMT was implemented. Following the loom shop, the blade shop application was described to us as probably the next watershed application. Although led by changes in the composite technology, there is little doubt that considerations of business and manufacturing strategy have played a vital role in the successful innovation. Clearly the decision to manufacture blades as 'sovereign parts' fitted an emerging business strategy of not manufacturing whole aircraft, just those parts with high added value and sovereign technology. Also, decisions not only to develop technology in-house to lay the composite, but also to plan and run the new shop using lead-edge manufacturing methods (JIT to reduce WIP), has contributed to overall cost reduction.

The 'Westland Crisis' could be seen as a significant emotional event for the Division. Such an event, which many believe is vital to stimulate change, can be either real or manufactured by management. Usually it constitutes a trauma which stimulates rapid and significant change within the company. The impending 'black hole' led to the 'Westland Crisis', the dramatic change in business context which precipitated a revolutionary shift in climate and culture within the company. This shift has resulted in Westland redefining itself as a company which continues to supply lead-edge technology, but across a much broader front of aerospace, whilst retaining control of the sovereign technology associated with helicopters. This is now the survival strategy of the company, and its impact can be seen on the very recent and highly successful AMT applications in the machine shops.

This FMS implementation came on-stream within this changed strategic and cultural context. The effect of the trauma has been the realisation, as has been stated before, that Westland is in business, not engineering. This has produced a much tighter structure and management style, and it has already been described how the FMS cells have been linked into the new business plan.

The restructuring (including the loss of 2000 of the workforce) which was forced on Westland might have several harmful implications. In order that labour reductions of such a high order should not impede restructuring, it is obviously vital to ensure that a 'critical mass' of those people able to contribute to and deal with the apex of a future business (prime contracting, systems integration, airframe assembly) should be retained, and that a 'red hot' procurement section is able to take care of the base of the business, where the bulk of the redundancies have come. There has not been a pro-rata shedding across all the levels. The company had to be left with the 'right people' who have the power to perform those tasks considered crucial to the business. However, a positive approach has been taken towards restructuring, which, according to one of the interviewees, followed the following plan:

Analysis $\longrightarrow$ Choice $\longrightarrow$ Criteria $\longrightarrow$ Action
(understand
each element
of the business)

This chain is extremely well illustrated in the FMS cell implementation, and is described fully in the text.

The company has now produced a five-year business plan; an activity which was barely noticeable in the 'technological culture' which prevailed in the past. The realisation of the changing nature of the business is exemplified in the equation:

$$COST = \text{Selling Price} - \text{Required Profit}$$

The business plan is enthusiastically expounded and promulgated by senior management. One person remarked that the atmosphere now was 'electric'. This excitement has obviously been communicated, encouraged and supported in the middle management teams connected with the blades shop and FMS cells. The importance of personal enthusiasm is well illustrated in both cases. The speed of the FMS cell implementation is remarkable. The business and organisational context was well defined, and this gives clear and measurable performance objectives to those working on projects. A significant emotional event triggered off a profound change within Westland. In this case, the SEE was a real one, but there is a contention, and an interesting one, that SEEs can be manufactured when further change is required. This fits in well with the idea of the 'performance ratchet' stage of any implementation methodology, which is a mechanism to keep alive the all-important excitement, electricity and enthusiasm, even when the worst is over, and a company appears to have turned the corner. Summarising the learning from Westland's:

1. The technology must provide business benefit, and this must be established as a first step. Interesting or novel engineering ideas are insufficient reason to invest.

2. Restructuring and retraining goes hand in hand with AMT implementation.

3. Following restructuring it is vital to get everyone on-board supporting the change. The lesson of the loom shop demonstrated this fully.

4. Top management must establish the contextual parameters governing (1) and (2) above, which establish the need for and scope of required changes.

5. Once the context is established and the support is there, rapid change, such as FMSs in the machine shop, can be implemented, but it is equally important to continue with planned improvement thereafter.

CHAPTER 6

Cummins Engine Company

The Cummins Engine Company site at Darlington is the subject of this case. It is one of three UK sites; as well as the engine assembly factory at Darlington, there are two other factories at Shotts and Daventry. The family of engines currently assembled at Darlington is in the medium horsepower section of the diesel engine market (i.e. 56 - 300 hp).

The original plant in Darlington was commissioned in 1964. This was as the result of a 50:50 joint venture between Cummins and Chrysler, who owned Dodge, and who wanted to recapture lost markets in the truck sector. There was a Dodge plant in Kew, but no room there to build this new facility. They were directed to Darlington, an area of high unemployment, but with a long-standing railway history and engineering skills. The railway engineering workshops there had employed 5000 at their peak.

The DTI and the local authority offered a site on the edge of town. The run-down of the railway engineering workshops coincided with the setting up of the new plant. Twenty two per cent of the capital was provided by the government. Cummins was to provide the

113

technical expertise and Chrysler the administrative and finance functions. 70% of the plant's output was planned to go to Chrysler, and 30% was to be sold elsewhere.

Cummins had a new concept in diesel engines. In the new Small Vee, engine size was reduced, and the alloys used instead of cast iron brought down the weight of the engine. A new building was added to the site to make the fuel systems, which were also unique. There were runs of 250, a JIT system was operating with three days' supply of materials, and the planned output of 120 per day, although never achieved, was closely approached: 107/110 per day being the highest.

However, sales of the Dodge truck never reached anticipated volumes and Chrysler could never take up its planned quota of engines from Cummins, who had themselves to absorb this extra output. They concentrated on selling to the North American market, producing engines for such diverse machinery as construction equipment, water pumps, agricultural machines, fire tenders etc. This diversity meant short runs, with the engine being tailor-made for the customer. Hence there were lots of parts. Under these circumstances 'JIT' was abandoned.

The peak in production of the Small Vee engine was reached in the mid-1970s. Cummins engines assembled at Darlington were sold everywhere. Ninety two per cent of their products were exported, but not extensively to Europe, where most manufacturers tended to make their own engines. Business with Mexico accounted for 60% of the demand.

In the late 1970s, at the time of the oil crisis, engine makers had to take a look at the products with fresh eyes. A fuel-efficient engine with a new kind of transmission was needed. There was investment put into the development of such an engine, for the Cummins strategy is to emphasise superior technological capability. The new type of engine resulting from this research became the B Series. At this time, Japan was beginning to get interested in building diesel engines.

114

In the early 1980s, the demand from Mexico, which had accounted for 30/40 engines a day, tailed off at the time of that country's own internal economic problems. This led to one Cummins plant in the North East was being closed down. The numbers on the Darlington site were reduced considerably, both in terms of directors and support staff. The Small Vee was reaching the end of its product life cycle and its successor, the B Series engine, was already being manufactured in Cummins' Rocky Mount, North Carolina plant, where it has been in production since 1983. It was designed in the early 1980s after the last oil crisis.

The decision to build the B Series engine in Europe, to serve that market as well as Africa and the Middle East, was made in August 1984, and the site chosen was Darlington. The site,

'... could have been anywhere in the world'.

The choice was supported by a three/four year deal with Leyland against a business background (1982-85) of uncertainty, a general run-down in the US economy, the Japanese presence coming on-stream, and massive over-capacity within the industry. It was decided in 1984 to build the engine on a Flexible Assembly System and this started producing engines in 1986.

The B series engine is lightweight, compact, and fuel-efficient, and offers, in comparison to its competitors, a reduction of 15-40% in parts volume and variety. This reduction has been achieved using the latest technology, assisted by 'state-of-the-art' CAD techniques, and contributes to the engine's relatively low weight and high reliability.

Although major components of the engine are still shipped in from the States, e.g. the base engine blocks, it made sense, in business terms, to assemble the engines in Europe for the European and neighbouring markets.

This entry of Cummins medium horse-power, in-line diesel engines into the European market opened up opportunities in the automotive, construction, power generation, defence and agricultural fields. At that stage this market was catered for by various well-established competitors, so the field was new to Cummins and in order to compete they, as a facility, had to concentrate on cost, quality and delivery in order to achieve competitive edge.

The site at Darlington is as technologically advanced as any in Europe, but it is the overall manufacturing philosophy/strategy into which this technology links which is seen as providing the guiding force for progress in the future. The maxim of 'cost, quality, delivery' is underpinned by sub-goals such as rapid response to customer demand, maximum flexibility (both in technology and manpower), first time quality, efficient materials handling (JIT is a feature of the work flow pattern), and maximum cost effectiveness.

The strategy pursued by Cummins in order to achieve these goals is seen by many as a 'back to basics' approach, which will be explored in more depth later in the report. Clear messages are communicated to the workforce about the company's position within the market, and what must be done to ensure success. There has also been a reorganisation of the employees into work groups which are affiliated to business units. This was seen as a necessary move from a more traditional structure to one which had a more business-like perspective.

There has been a recent history of heavy redundancies both in Cummins itself and in the surrounding area. The Cummins plant at Darlington had two sites, one making components for world-wide distribution, and the Cummins B Series site which also contained the Small Vee facility, which was gradually being run down. 1979 was a peak year for employment, and between 1980 and mid-1983 there were only seven redundancies announced site-wide. Since then, the workforce has declined across the whole site from 2800 to 1000, and the introduction of the largely automated B Series facility saw reductions in that area from 600 to 140.

The B Series facility has the capacity to produce 40,000 engines per year. The shortest engine build time from door to door which is capable of being achieved is one shift, i.e. just seven and a half hours. At present, the build time is three days, so there is obviously a shortfall between potentiality and actuality. However, an equivalent North Carolina facility takes considerably longer and, as one manager observed, the build time of 3 days had been preserved even though the engine output per day has quadrupled, from 20 or so, to 80/85 per day. This has also been achieved without a corresponding increase in manpower. The approximate present figure of one man/engine/shift was given. The ultimate target is 0.8 man/engine/shift.

The facility had been planned to cater for a medium-high volume/low customer number market. This plan was in the face of decreasing market volume, decreasing customer base, and finite product life. No longer was it possible to predict a product life cycle of 15-20 years (as it had been for the Small Vee series). The life of the new engine series would probably be less than half that. Any company competing in this market would have to be able to address these areas effectively. Part of this ability was seen as being provided by the built-in flexibility of the new facility. However, the market had turned out to be rather different. Instead of large numbers being supplied to few customers, they had orders from an increased customer base for hundreds of different engine specifications in less than 18 months. An example was given of the June 1987 build which was: 135 different specifications working out at about six per order. Here, flexibility still served them well; in fact, it was remarked that it was needed more than ever.

The Technology

The Factory

When the decision came to build the B Series at Darlington in May 1984, the area which had been used to manufacture the Small Vee engine was converted within 13 months, which was seen as an impressive achievement by both participants and professional commentators alike. The Small Vee factory was 20 years old and run down.

> 'A dirty area'
> 'a horrible job ...'

were just two of the comments. The floor space devoted to the Flexible Assembly System for building the B series is 150,000 square feet, and has the appearance of being clean and new. Areas set aside for team meetings have noticeboards covered with neat notices about rotas, training schedules, etc. This is very reminiscent of those workplaces in Japan reported on by Schonberger [see Schonberger 1986].

Work began in late 1984 and the first engine was produced in May 1986. The layout of the shop-floor divides the engine build process into discrete units, itself an aid to flexibility (different operations can clearly be directed to the appropriate area) and there is a simplified workflow process throughout the facility. Great attention was paid to the layout planning and this has had knock-on effects of simplifying workflow and movement of materials on the shop-floor. Cummins had experimented for years with various socio-technical designs, from flowline, to ride-the-track, to teamworking.

The Flexible Assembly System (FAS) is laid out according to the following plan:

BASE ENGINE	Short block assembly: Inclusion of main engine components such as crankshafts, camshafts and pistons.
	Long block assembly: Addition of cylinder head valve train and front end gear train to the block.
	AGV Upfit: The engine at this stage is a basic assembly and it is during this next phase that different specifications are introduced, depending on whether engines are to be naturally aspirated or turbo-charged, and what form the fuel system will take. This phase of engine build is carried out on automated guided vehicles.

UPFIT

| Engine Test: Pressure testing of oil, water and fuel seals, after which the
| engine is subjected to hot test, where it is connected to all major services
| via quick hook-up couplings, and transferred by computerised rail car to
| one of three test cells. Here, in under six minutes, engines are given a full
| and exhaustive test run.
|
| Post-test trim: Accepted engines (rejected ones having been re-routed)
| now have customised items added - again being transported by a second
| set of AGVs. This is sometimes referred to as customisation upfit. Fly
| wheels and their housings, compressors, power steering pumps and so forth
| are added to customer specification.

TRIM

| Engine paint: In this area, the engines, having been transferred to an
| overhead monorail conveyor, are carried through the computerised paint
| system, comprising phosphate pre-treatment and choice of 12 colours.
|
| Post paint trim: In this final stage, items which cannot be painted are
| added, e.g. electrical components. Other customer options are included,
| such as radiators, gear boxes, clutches, transmissions, etc.

The completed engines are taken from the conveyor and placed on an overhead monorail ready for final mounting on engine skids prior to despatch. Receiving and despatch bays are so arranged that, in combination with the JIT simplified materials handling in operation throughout the shop, a component for engine assembly is either,

- at the receiving bay

- at the point of use (assembly)

- or (briefly) en route between those two locations

- or at the component machining area eg. Flywheels

Although seven distinct areas have been identified on the shop-floor, these fall into three identifiable engine build environments: base engine, upfit and test, and trim/final assembly, which enables rapid responses to customer demand to be made. There is a quick, clear route for each engine to follow.

The Equipment

Major sections of the FAS were turnkey, and the layout of the shop-floor, having been described the equipment is here explained more thoroughly.

In the long and short block assembly areas a power and free conveyor system supplied by Krause of Bremen is used. This is an elevated non-synchronous conveyor system incorporating manual and automatic assembly and sub-assembly operations, and was required to provide a rigid base for automation.

In designing the system the level of automation had been justified. It was felt that, as far as Cummins was concerned, the volume did not warrant full automation, and such automation as had been introduced had been tailored to its own volumes.

Up to the end of long block assembly the engine was fairly standard, although, as people working in the area commented, 'standard' is perhaps a difficult word to use,

'... but every engine's got pistons - although it might be a set of four or six.'

Before passing on to the next stage, the engine was equipped with its cylinder head, bottom and top ends complete,

'.... with no real peripheral bits added'.

Up to that point too, many critical operations in terms of engine build are carried out. These include torquing up the main bearing caps, torquing up the cylinder head etc., all operations which will eventually,

'... dictate whether or not you've got a good engine. We felt we could automate some of these critical operations, and this would be easier to do on a conveyor track which automatically locates the engine.'

The cylinder head sub-assembly is totally manual because automation of this operation would have been quite expensive, and is,

'... not justified by our volumes'.

Therefore, it was decided in this long/short block area to go for automation, whenever there was a critical joint or a critical quality check.

It is in the next area, the AGV Upfit, and in a subsequent one, the Post-test trim, that the automated guided vehicles provided by Jungheinrich are used. At the end of the assembly,

'... we will more or less do what any customer wants us to do - within a very wide range. This includes a large variety in both the location and type of parts. We knew we needed a very flexible system.'

Cummins had developed the ideas over some considerable time.

Looking at options had started some time ago, and not only in connection with the B Series. There was,

'a feeling in Cummins that building engines on a moving track was not the thing to do, so we started looking at all sorts of methods of building engines, none of which got off the ground. We never had any money to spend, and volumes were reducing.'

A small amount of research had been carried out on changing the existing Small Vee methods, but ideas were always, 'put on the back burner'.

However, when the decision to make the B Series at Darlington was taken, these ideas were,

'all dug out again, (for) that was the way the technology was going'.

But this method could not be adapted to making anything else, and the new system, had to take on board the fact that engine life would be less, and that new products might need to be made on it. Not only had the new system to possess design flexibility (be able, for instance, to build an eight-cylinder engine) but it ought also to be capable of adapting to volume flexibility, to have the flexibility to respond to customers' 'add on' requirements.

Many options were considered, including that of extending the track which they already had. Finally, the decision was taken to use AGVs, 16 in the 'AGV Upfit' area, and 10 in the 'Post-test Upfit' area. It is expected that a total of 40 AGVs will be used once a second development phase has been completed. Following a guide wire inserted in the factory floor, the AGVs are equipped with a special hydraulic lift table which can be adjusted to the most convenient height for the individual operator working on different stages of the engine assembly. The track is laid out in the sequence in which the engine is built.

'We had the intention that it is fairly flexible to alter if you get changes. You can cut a slot in the road and lay a new wire. This would be much more difficult with conveyor tracks.'

The engine is kept track of from build start by a sophisticated management information system (AIMS - see later) so that when a block goes into station, its bar code is read off the computerised pallet on which it travels, which sets the machine into motion for the right configuration of engine.

'All along the line there are programmable machines, so that if someone comes along in the future and says 'we want an eight-cylinder engine', we just reprogramme the machines and we're in business.'

At the end of the long block assembly line, the relevant data concerning the engine are communicated via AIMS to a computerised data convertor on-board the AGV. This provides the routing instructions through the Upfit areas, so that the appropriate fuel systems etc. are built. An innovative feature of Darlington's use of its AGVs is that engine build takes place on the AGV, the operator travelling round with it. Usually, this kind of vehicle is used for bringing material to and from central stores or to and from workstations.

The loops through which the AGVs travel, and the stations within a loop at which they are programmed to stop, depend on the engine build. The operator with the appropriate kit of parts travels with the AGV, assembling the relevant components on the engine as it travels through the loop. The speed of the AGV was initially 2.4m/min but was increased later to approximately 3m/min. The AGV takes 12 minutes to pass through one loop. Parts required in the Upfit area are transported to zones adjacent to this assembly section, and a list is picked for each individual engine from information generated by AIMS and displayed locally on VDUs.

If engines are accepted in the quality audit area, they then proceed to the next stage. If they are rejected, they can be re-routed automatically on their AGVs to a quality maintenance layby where minor faults are rectified. Engines with major faults are stripped down and their parts redistributed back into the system.

The next section is the hot test area with equipment provided by Schenk where the major part of the test routine is carried out in one of three automated test cells through which the engine is carried on a computerised railcar. This is one area which continues to give problems. The railcar is fully computerised and the serial transmission of information is across a number of control system interfaces.

> '... something's going wrong here. To de-bug that equipment we need a highly-skilled team, which we haven't got. The railcar still goes loopy, even after 14 months.'

When a system relies on this type of serial communication in order for something to happen, one person remarked, then if there are problems, sorting them out can be difficult. This, in his view, was an illustration in miniature, of the problems of totally computerised systems, and the questions of level of control which ought to be addressed when planning any new facility which incorporated elements of computerisation and automation.

The final area is engine paint, which has an automated painting system with robot paint applicators.

Management Information Systems

MRP has been in place for six or seven years. Information from this system, i.e. what parts are available, what status they have, how many there are, etc. is passed onto AIMS (Assembly Information Management Systems - a Honeywell system). This passes information via pallets upon which the engines rest and bar codes on the engines themselves, directly to the equipment and personnel responsible for the production of the engines. It outlines the build requirements and specification of each engine that needs to be built that day and monitors each engine as it passes through the various manufacturing stages from the time that a pallet becomes available on the short block line, until despatch after post-paint trim. It was stressed that this was a prompting and recording system, not a control system.

> 'AIMS takes away the paperwork and all the mundane stuff, but not control. AIMS could fall down and we could still carry on.'

Information is stored on electronic memories on control tags running underneath the conveyors. Serial numbers of each engine are recorded so that parts can be tracked if there are any problems. On completion of each assembly stage, an assembly computer reports back to AIMS on the completion and status of each operation. This information is then passed back interactively from the Honeywell host computer to the mainframe business systems, so that ordering and material control systems can be updated.

There are sometimes problems with some of the parts, e.g. the fuel pumps. MRP might signal for instance that these parts are unavailable at the start of engine build, but often a conscious decision is made to allow the engine to go ahead because the fuel pumps would probably be available later.

'MRP tells us 'no', but we cheat and AIMS is instructed to go ahead.'

This was given as an illustration by one person of the flexibility of control which remained with the operatives.

Purchasing, Materials Supply/Planning, Materials Flow, Materials Handling

Although not strictly part of Advanced Manufacturing Technology, the way in which Cummins deals with these aspects of the business are seen, to quote from their brochure, as an 'Advanced Manufacturing Methodology', which also includes such areas as quality assurance and working practices. However, in the matter of inventory control purchasing is carried out electronically via MRP, and there are no goods inward or central storage facilities. These have been replaced by a combination of on-line bulk issue in vendor packs and option component kitting. Supplies of materials and components are therefore delivered on the day they are required on a JIT basis, and go directly to the point nearest where they will be assembled. It was felt that there had been some hiccups in this, but by and large the

JIT workflow within the facility is perceived to be working extremely well.

People in Cummins had been,

'thinking JIT' for a long time. Someone comes along and dresses it up in a fancy name, but it isn't a new concept'.

However, its formalisation by way of the well-known acronym had made them look at it from a different perspective. They had been doing it, but now there was a formalised way of doing it.

'It has helped; it does make things more effective.'

JIT had been instrumental in determining the layout, the manning and the control in the new facility. Cummins had been helped in this by the fact that there was a strong industrial engineering input from senior management. JIT was perceived as being very useful in capital planning.

It was realised that with many of their major components arriving from Cummins in the US (e.g. the engine blocks) and some coming from Japan, a JIT system could not properly be said to exist. Having materials in a 2000 mile, or even longer, pipeline, was a contradiction. However, there could be no justification in building, for example, a block-producing facility next door. Cummins UK did not have the money. On the other hand, Cummins US had the capacity.

However, efforts were being made to,

'get everything right that we can get right'.

JIT has implications for,

- Suppliers. Initially these had been reduced from 350 to 80 although, later, numbers began rising again due to proliferation of parts. Parts were single-sourced; their volumes were not big enough to make anything other than single-sourcing economic.

- Locality. Many items were now being made in-house, or resourced locally. Wherever supply chains could be shortened, this was being done.

- Frequent deliveries.

- Small batches.

- Standardised containers, although there had been problems here. Initially consultants had recommended the use of standard pallet sizes based on the quantity of US-sourced material at the start up of the plant. This is not compatible with European standards as the supply base changes over time.

- Quality. All suppliers are quality audited with Quality Agreements entered into and reviewed at regular intervals. Critical operations in the manufacturing process are automated. Statistical process control is being introduced as a quality control tool to the workforce. At present SPC only takes place in the automatic machines. There is a core of people being trained in SPC in some of the manual operations at the moment.

Purchasing/materials supply people had been involved right from the beginning. The positions of set-down areas, containers, packaging, etc. had been their decisions when it came to layout. It is perhaps a measure of the success of JIT that the receiving area was very much larger than it actually needs to be. In the planning stage, it was not clear how much space would be needed for handling material, making quality checks, etc. One person thought that facilities of storage of raw materials could be planned on a need basis rather than on a planning basis. These facilities are quick to install. It was felt that there would never be a perfect JIT system, just a closer and closer approach to one.

The Rationale and Justification

Some discussion on the technical rationale of the AMT has already been presented, for instance: where and where not to automate, development of suppliers to fit in with JIT, etc. What follows here is a broader view of the reasoning behind the setting-up of a Flexible Assembly System. In total, there has been a £13.5m investment in new plant and equipment (the computerised railcars alone cost £2m) and a total investment of £40m in the B Series facility.

Initial Plans

In late 1983/early 1984, Cummins UK was asked to make proposals to the main board of Cummins Engine Company Inc. in the US for a plant for assembling the B Series engines in the UK. As already mentioned in the 'Background' section, it made more sense, in business terms, to make and distribute the engine to European and neighbouring markets from a manufacturing source within Europe.

A series of proposals were put forward which were dependent upon the volume of engines expected to be produced by any new plant which might be built. High volume would dictate dedicated lines, medium volume a Flexible Assembly system, and low volume an assembly line.

The person chiefly involved in these calculations had been in charge of much of the AMT planning across the UK, but a number of people were involved in preparing the package of documentation (PAM) which went forward to the main board in the US. It was he who edited the proposals of the PAM team, consisting of one manager and two or three engineers. Resources and expertise from Shotts, Daventry and Darlington were pooled during the preparation of the justification, which was then subjected to several iterations before presentation to the main board. The PAM preparation and finalisation was regarded as,

'the hard bit - essentially this is getting the politics bit out of the way - after this, we can just go ahead'.

Financial Justification

Great importance was placed on the financial justification of the proposal, given the view that this should not be,

'a blind leap in the dark', or 'an act of faith'.

It was realised that many AMT justifications fell into just this trap, and that whilst it was recognised that cost benefits of such installations fell into two categories,

'the tangibles - the easy bits as far as the accountants are concerned, and the intangibles - much more difficult',

the attempt still had to be made to quantify those intangibles. The Manufacturing Services Director had written a paper on this very subject. It contains a worked example to show how this quantification might be carried out.

Built-in Flexibility

Cost, quality, and delivery (see later) was a policy forming the base upon which the rationale for the FAS was grounded. This base was seen as being best served by a built-in flexibility, which was enhanced by JIT and the design of the assembly system itself. This flexibility has already been referred to in earlier sections, and is seen as crucial in business terms where (a) Cummins is entering a new market, and (b) where the product life cycle will probably be halved or more in the near future. The ability to change its assembly methods to accommodate changes in volume, type of product and/or parts which could not even be predicted at the moment, would be vital in such a fluid market. Flexibility is also vital for the type of customer which Cummins is now attracting, with their demands for high product variability.

How? Installation and Implementation

Permission to go ahead on the FAS now in place at Darlington came in August 1984 from the US main board. The B Series facility was erected on the site of the old Small Vee engine factory. From the evacuation of this plant, cleaning it, getting in the civil engineers, refurbishment, machine installation to production start-up, took thirteen months, which was only three weeks later than planned, and there was a slight underspend,

'... a bit of a surprise in the States!'

The first B Series came off the line in May 1986. Within this period were included six months spent in sorting out the details, and a further six months between ordering the machines, and installing them.

The Project Team

In charge of the whole project was an engineer who was described as 'relatively independent'. The sorting out of details mentioned above was going on at the same time as capital purchase, but the acceptance by the main board and the spreading down of the news of this acceptance throughout the rest of the organisation, was seen as having removed, as referred to earlier,

'... all the politics from the business'.

Decisions on purchasing particular parts of the installation became almost the same as the purchase of a standard machine tool, albeit on a much bigger scale, involving more people,

'... but nothing special'.

It was realised in planning out the critical path for such an installation, that something of the order of 1,800 discrete events were involved.

'There were so many different people in so many different areas, with different responsibilities .... the critical path network had to be right.'

Planning was very much assisted by the presence of a strong industrial engineering base at Cummins. Various managers were called in at successive stages of the installation, as and when their expertise was needed. There was an overlap between people who had helped in the preparation of the PAM documentation, and those who were directly involved in project management, and throughout the whole process, central group (i.e. not plant-related, e.g. Daventry, Shotts and Darlington, but offering a resource, particularly advice on systems) acted as internal consultants.

The engineer who was the project planner eventually had a team of managers, each of whom was responsible for one aspect of the FAS, and each of whom had expertise/experience in that area. Thus the manager in charge of the AGVs had been involved in the successful installation of AGVs elsewhere. The only exception was the test area, which was the most complicated, where two managers were involved. These were the people who,

'went out and got the quotations in the first place, and who 'developed' the suppliers'.

Different departments had various perceptions; for example, some thought that the project had been managed,

'.... in a participative way,'

whereas another view was that,

'... the systems were working, and my people contributed to the plan'.

As the machines began to be installed, business managers (see later) and some of the workforce, were able to go in and start to look around.

'We found one or two obvious problems, but these were soon put right.'

The project planner and his team were under the overall view of the project director, who had,

'more of a global view'.

The different parts of the project were subjected to the kinds of iteration already referred to with respect to the preparation of the PAM documentation. There was close liaison throughout between the project director, project planner, the people responsible for the layout

132

of the new shop-floor and the machine tool suppliers and installers to make sure,

' ... that the whole thing meshed together .... came together at the end, and it did.'

The Suppliers

Each section of the FAS was put out to competitive tender, and each tendering company was rigorously checked both with respect to the functional specification of their turnkeys, and to their financial position. In this, they were helped by consultants who gave advice on the functional specifications required for the kind of FAS which was being envisaged,

'... although there was lots of in-house expertise too,'

and who also helped them on aspects of project management, and the management of contracts with suppliers.

At least two of the suppliers of each of the major technologies were identified, and a quotation from each was requested. The project director visited those who seemed to be the best suppliers on paper, but at least two of the submitting companies were developed right up until point-of-sale, in order to maintain maximum bargaining power in contracting.

After the tenders had been awarded it was still necessary, in the view of some, to keep the successful ones on their toes. An example was quoted of one of their suppliers going through some organisational turmoil during the time of their contract with Cummins. The Managing Director of the supplying company had been replaced, and there was slippage. As soon as this was noticed, Cummins followed this through, and the suppliers were told, in no uncertain terms, that,

' ... this just isn't good enough.'

The situation improved but,

' .... it's no use expecting them to meet deadlines unless you're pushing them. You come across this sort of thing everywhere, people promising things, and then the dates slipping. Our time-frame seemed short, but we stuck to it.'

Some of the suppliers are still on-site, either providing some input on technical training, or sorting out some of the technical difficulties which are encountered from time to time.

The Consultants

It was the view of one of the participants that,

'consultants rarely give us anything we can't do ourselves,'

but it was felt by more than one that the consultants who had given them advice on developing the functional specifications for the FAS and had provided training on project management, had been good.

'We were well served'

was one of the comments made.

They were perceived as having had lots of management experience; that the formal documentation they used was good and their formalised methods of procurement did work. Although,

'.... the way they do it makes the project expensive they take in a goodly per centage as risk money in case anything goes wrong ......',

the consultants were perceived in this case as having provided value for money. The ability that Cummins now has to use the consultants' form of contract documents when ordering/purchasing equipment in excess of £500,000 was also a useful spin-off.

Control

It was the view of some that the time and effort devoted to determining the level of control within the system had contributed to the success of the facility which they now had, and which enabled them to operate with the flexibility which they did.

'We've driven control down as far as we can get it.'

'We've chosen the best control systems (Siemens) but everything's an independent unit ..... there's very little communication where there doesn't need to be, therefore there's less to go wrong. It's a lot easier to keep a programme in one computer, rather than in ten.'

'As far as possible, everything operates as an independent unit.'

The rationale behind this simplicity of control was that it enhanced flexibility and therefore response to the market place. Examples were quoted from the US engine plant at Rocky Mount, N. Carolina, where there was much more computer control, and if there was a problem the whole production line could be halted. Computer 'lock-ups' of this nature could be disastrous. Rocky Mount's engine build time is 12 days as compared with Darlington's three days. The problems of sorting out serial transmission of information in a fully computerised system were illustrated in their own hot test computerised railcar system.

'Imagine if we were having the same problems with the whole of this facility.'

In summary then, much time and effort was devoted to control systems, in the broad sense of the word, rather than just to narrow computer systems. According to one person, this insistence on simplicity, upon a system tailored for them, had paid off, although it had involved a lot of time at the front end.

'Once you've sorted out the organisation, sorted out its controls, you're beginning to understand things .... if you can maintain production when sophisticated systems break down, then you're halfway to running a successful enterprise ... we have spent (on another site) between two and four years on the control systems we want to prevail, in order to give us the simplicity we want.'

It was felt that once the materials movement side of things had been sorted out, then controls became far simpler, much simpler than those which would have been recommended by their turnkey suppliers. The interviewees were aware that they themselves still had some way to go as far as materials movement was concerned.

Organisational Culture

In our view, the organisational culture of Cummins provided a strong supportive context for the implementation of AMT. Rather than analyse this culture in detail, it is illustrated below by Cummins' approach to communicating their corporate philosophy, and the way they have recently tackled a company-wide programme to reduce costs. Both of these examples illustrate well what we regard to be the major strength of the Cummins culture, namely its (as Peters & Waterman have termed it) simultaneous 'loose-tight' properties. By this they mean the coexistence of firm central direction and maximum individual autonomy. The 'loose' qualities of this culture encourage innovation and 'back-to-basics thinking' on the issues facing the company, and in particular the implementation of AMT, whilst the 'tight' qualities ensure that change is tied into overall business and corporate direction.

The Cummins culture was variously described as 'open', 'easy' and 'participative'. This is perceived to stem, in no small part, from the US parent company, who positively foster this kind of atmosphere.

The participative nature of the company is dependent upon a well thought out communications strategy, whereby the core values and philosophy of Cummins, as an organisation, are spread to all its members. The messages thus communicated are short, succinct and easily remembered, and both their content and promulgation are enthusiastically expounded and supported right from the top of the organisation. This has the effect of providing firm direction from the centre, but encourages maximum discretion in implementation.

Communicating the Cummins Philosophy: The Cascade

'We complain about communications, but we're good at it'

captures the essence of Cummins' communications. Messages about:

- who they are
- what business they're in
- what the market is like

are transmitted both formally and informally. Every quarter, there is a video presentation sent by the Chief Executive in the US. This has been happening for the last five years. In the beginning, according to one person,

'... the message seemed a bit fragmented, but as the years have gone by, it's come together for me, and I'm sure for others too. That helps in management, the Plant Manager, us, everybody. The video gets over what the corporation wants and where we stand exactly.'

There was a newsletter from the Plant Manager and a slide show presentation once a month.

'This tells us about the plant, the customers, what's happening internally.'

The work groups meet once a day for about ten minutes, and the Business Managers are,

'... always walking around, talking. The very way we work, in teams, means that we're communicating all the time.'

Ideas, as well as proceeding from the top, are generated at seminars and teach-ins. There are annual planning conferences at a corporate level in the UK,

'... whose findings are passed up to the top and then get spread down again.'

There were many other techniques referred to for 'spreading the message down' or cascading the company values. Reference to one such technique will perhaps serve to illustrate Cummins' way of operating in this area. A package, referred to as NSE (New Standards of Excellence - again it was the acronym by which the package was known; a sign perhaps of having passed into company vocabulary) has been developed in the US, by training teams in Columbus. This package, comprising NSE 1 and NSE 2, has been slightly reinterpreted for the British sites, and it addresses,

- the way people work
- the way people manage their work.

This package was available for use by the British sites, and the Darlington site chose to adopt aspects of it to address specific issues.

NSE incorporates a number of what were referred to as models. For example, one model shows the company as a traditional organisation with instructions going down from the top, through the hierarchical levels, to the shop-floor, and information passing from the shop-floor back up to management,

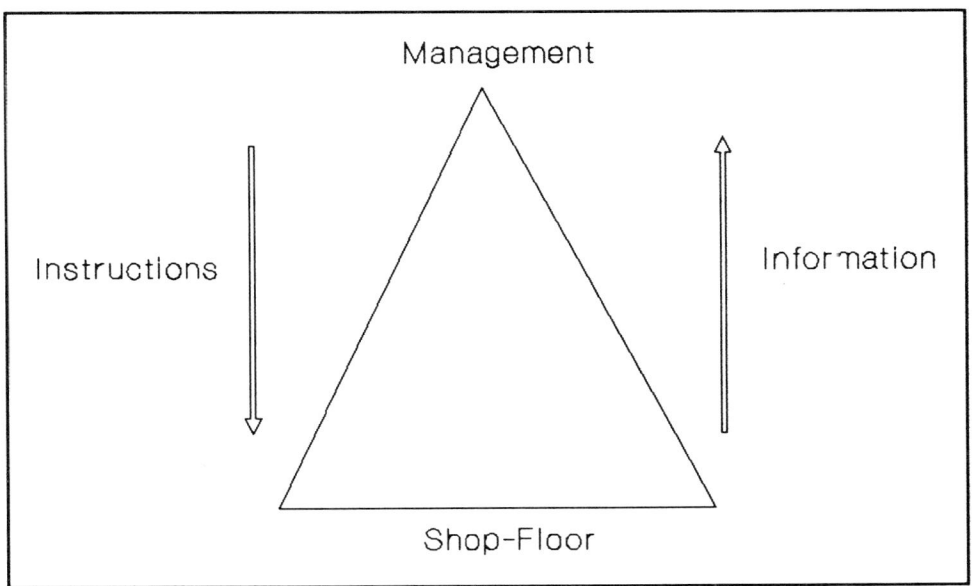

But the company could, according to the model, be turned upside down, and this view of an organisation could be equally valid, particularly in the phase in which Cummins finds itself now,

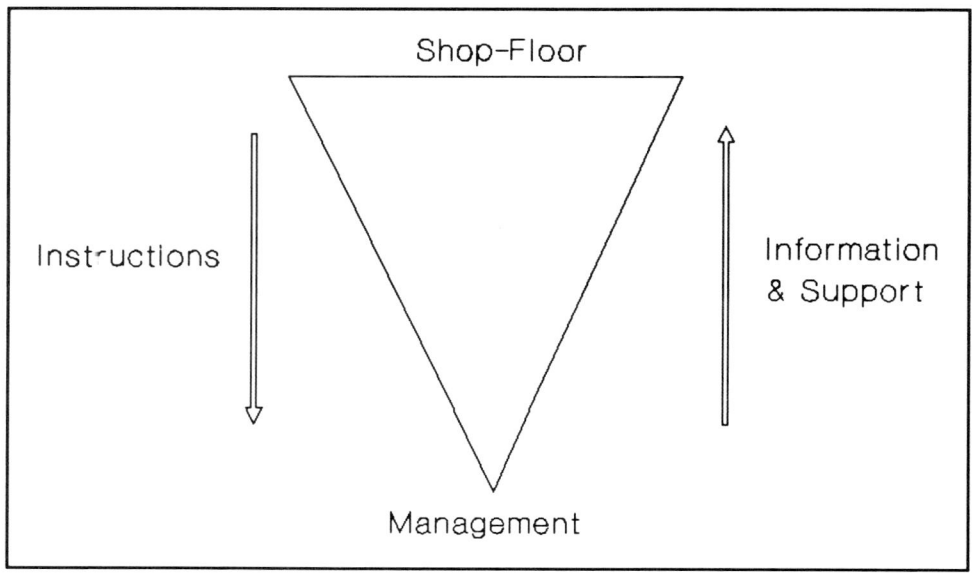

This reversal stresses the fact that it is the shop-floor,

'...... the work groups, that get metal through the place and engines out of the door'.

There were nine different models in this particular package, all aimed mainly at supporting the work groups. Another model is of a long face surrounded by 5 smiling ones, as a reminder that problems should be shared, not kept by one person,

'.... talk about it'.

There was a feeling that the work groups had been told of NSE by management, but had not really yet been given an opportunity to learn it. One person remarked,

'We've got the package (NSE) now, and we've got to lift people up another step, so that they understand what the company is trying to do ...... it's not the manager managing, but everyone managing the way they work'.

There was a feeling that there was now, as a result of these kinds of communication and the ways in which they had been carried out, a much more realistic appreciation of the market environment they were in,

'.... by everyone, management, shop-floor, Unions, everyone.'

People, it was said, were now accepting that they were part of the whole drive towards CQD (Cost, Quality, Delivery - see later) and that the company were looking for more than a nine to five commitment, and that the way in which each person 'managed himself/herself' was crucial to the enterprise. A graphic illustration was given of Cummins, in a way expecting the quality of self-management which people quite normally exhibited in their home environment.

'When people decorate a room, they do all sorts of things - they cost the job out, they're careful with quality. They don't have to be told what to do, and they do a good job. Why should they be different as soon as they step over Cummins' threshold? Management used to be happy to just come in and do a day's work, but it's a different business now. I was as bad as anyone else. On the Small Vee I thought all you had to do was to keep making them - they could be dropping into a black hole as far as I was concerned. Now we know that we've got to satisfy the customer, otherwise he won't come back for another engine..'

It was perceived that much effort had been devoted to changing the culture on the shop-floor, but initiatives were being directed at management too. There had been a move to widen the manager's span of control. This had been going on for some time.

'.... Trying to get people to take on other responsibilities, to get the job done, you don't have to wait for someone else, just go on and do the job yourself'.

Taking Action and Delivering Results: Cost, Quality, Delivery, and the Sprint

During the period through which the FAS at Darlington was being designed and implemented, the parent company in the USA had launched what they termed the 30-month Sprint. In face of increasing international competition, Cummins had found themselves losing market share.

The US main board had issued a firm directive that costs in the organisation must be held at the present level for the next 30 months. Competitors' costs, especially those in Japan, were lower than those in Cummins. How this requirement was going to be achieved was left to the initiative of top management in the various UK subsidiaries.

The UK board launched a major drive at improving cost, quality and delivery (CQD) as the central plank for achieving the goals of the 30-month Sprint in the UK. The CQD message was cascaded down the UK organisation and supported by widespread training, as well as a series of initiatives described below. Cummins Engine Company in Darlington was faced with a stark choice, either to compete head-on with the Japanese, or close the factory. Their own costs were 15% higher than those of the Japanese and, within 30 months, they would have to get rid of this differential.

The acronym CQD was on everyone's lips, and was seen as a policy which provided the firm behaviourial foundations for the success of the Flexible Assembly System. 'It increased awareness of the need to change, and it created a vehicle for change.' The content of the message was COST, QUALITY and DELIVERY. The SPRINT was described as creating a focus for the need to change. The NSE message described in the previous section could be seen as the ratcheting up of the workforce's performance and keeping it to the clear objectives of CQD,

> '..... NSE is to lift them up to the next level'

an incremental step. CQD, delivered by the 30-month SPRINT, was described as,

> 'a quantum leap'.

Initial reactions to the message and time-scale were vigorous.

> 'People were saying 'it's impossible', but it worked. There was a dramatic reduction in our cost base.'

It was observed by some that even if they had only achieved 60% of what they had set out to achieve, the exercise had broken the inertia, so that the company could,

> '..... move forward into an improvement mode.'

JIT was seen as the only way to proceed in order to reap full advantage of a Flexible Assembly System, but one person remarked that it was still very hard,

> '..... getting quite simple ideas across to middle managers ..... the fact that JIT means Total Quality Control (TQC) as well ..... I suppose though, that this involves a fundamental shift in most people's thinking, the fact that if you get the suppliers right, and the materials right, then you get the process right.'

The Japanese incredulity about the Western practice of quality control was also referred to.

> 'Before we can do these things properly, we must get our own house in order first.'

Quality and delivery are enhanced by JIT, and quality is being further improved by the introduction of the workforce to the ideas of SPC (statistical process control). JIT also reduced costs by cutting down on costly inventory and WIP. An incentive to reduce costs was provided by the example of 'Third World' countries, with their very much lower labour costs.

There was an observation that both the Sprint and CQD had, to some extent, been seen as something of an American,

> 'flavour-of-the-month, but it was more than that; it was revolutionary.' There were similar feelings about JIT, TQC, and other things, but they too have been revolutionary. It's no good people just putting their heads down and hoping it will all disappear; it won't.'

Harmonisation in Cummins: the Implications for AMT Implementation

Background

In 1984, when thinking about the future needs of the company, it became apparent that alterations in certain areas, working practices, trade union negotiations, work organisation, etc. would be required, not only in Darlington, but UK-wide, including the sites at Shotts and Daventry.

Not only was a new product being introduced in Darlington with the accompanying massive cash injection into an FAS to support a new manufacturing process, but FMSs were being planned at the other two sites, which would have similar organisational impacts.

In spite of Cummins' above-average levels of flexibility and skills, it was perceived that

144

historically, these had not been used to their full extent, and that they would need to be brought out in the coming testing period of the company's life.

Against this background, and with the need to actualise the potential skills and flexibility within the workforce, there were two major thrusts within the personnel function:-

1. To get the support systems right in order to encourage that actualization.

2. To address specific training issues which would be needed with the re-designed structures.

Support systems: the Trade Unions and New Pay Structures

It had become obvious during the period of rapid company contraction that having to deal with several unions (two in Darlington - the AEU and ASTMS; four in Shotts) would hinder the development of the flexible working practices needed to support the introduction of the new product, and the advanced technology associated with that introduction.

> 'We didn't know it at the time, but we were a JIT delivery system. We were having to change people from the block line in the morning to making crank shafts in the afternoon. In the old days, we'd have had two separate crews, but contraction meant that we were going to have to move people around like that all the time - as a matter of course.'

This produced personal stress, and although 'flexibility' had been built into work agreements,

> '....we hadn't given them the skills to be able to be so flexible.'

This personal stress and the negative reactions against the disturbance of normal working patterns resulted in the company resolving to follow a positive course of action which would overcome these difficulties - harmonisation across the sites, and a training programme which would provide incentives for flexibility.

The Trade Unions

The ASTMS negotiated on behalf of the staff and there were ten pay ranges in all, starting with routine clerical workers, security and catering staff at one end, through supervisors to the professional grades such as metallurgists and accountants. The AEU had six grades in all, two at craftsman (skilled), three at semi-skilled and one at unskilled level.

The objective at Darlington was to integrate these two bodies and to 'crunch down' the number of grades. This has been done and Cummins is now a 'single status' organisation with five broad pay bands applicable across the board, whether to operatives working on a diesel engine or in an office, thus integrating previously separate blue and white collar jobs. This did away with the complex bargaining structures, the separate job-evaluation methods and the different grading and payment systems.

Not surprisingly, this harmonisation involved difficult and protracted discussions lasting six or seven months across the whole of the UK. The major difficulty was in persuading people that the new arrangements were not threatening, which was hard, because at that time Cummins was going through a period of contraction. Negotiations went as high as national level officers.

'We had to be seen as even-handed and fair....to try and convince them that we were doing something forward-thinking and progressive.....'

Once agreement on merger had been obtained in principle, the detailed work of integrating two quite separate pay systems and union structures was accomplished,

'..in a surprisingly short space of time.'

Management decided on job allocations based on their previous relativities, and then discussed the bases of these allocations with the unions,

'We were able, by continuously talking to each other, to avoid a massively expensive job-evaluation exercise — which was reckoned to be quite an achievement. We got surprisingly few problem areas....we were able to reach a conclusion pretty quickly.'

Models of Harmonisation

Consultant, University and Business School help had been sought in devising job-evaluation/payment systems etc. but,

'....no-one seemed to have thought of a system which could cope with the range of things we were dealing with. We were generally disappointed. Groups known to be interested in this area just couldn't cope with the urgency of what we had to do; they (the suggested schemes) couldn't be applied in any quick way. We had to be ready when the B-Series was starting in 1985. This (the manufacture of the new engine at Darlington) hadn't been agreed technically but, internally, we were pretty sure that it was going to come here. We researched the area ourselves, talked to others, did a lot of reading.'

The model which was eventually 'hand-knitted' by Cummins themselves, was thought to be forward-looking and innovative. Issues and problems have undoubtedly cropped up, but the harmonisation has been a key part in gaining employees' cooperation on training, and the development cf the new skills necessary to support the new technology. Internal equity,

'....we were looking for natural justice within the community of Cummins'

had been preserved.

The harmonisation was considered to be,

'absolutely fundamental',

and had to be well in place,

'...or at least all the markers had to be down before we even went out on the shop-floor.'

The process was regarded as a central part of Cummins' strategy and, had it not been,

'.....tackled at the front end....we certainly wouldn't have had the implementation going so well.'

Pay Structures

There are five broad pay bands across the UK although, in practice, the bottom pay band, (reserved for employees who could not train for some reason or other eg. disability), contains only about six people at Darlington, and about twenty UK-wide. Progression through the pay bands is achieved by acquiring new skills, and it is this incentive which is seen as the bed-rock upon which the attainment of extra flexibility rests; hence the fundamental nature of the re-design.

Training Issues

'Technology was moving at such a pace, and we were going to be introducing new products and new manufacturing processes, that whatever we'd done before on training and development was going to be knocked sideways.'

The harmonisation had produced the right background for radical ways of looking at training. Narrow pay bands would have restricted employees' acceptance of acquiring and applying new skills. Now there is a much bigger population in each group and more motivation for people to extend their skills which have to be continuously updated. Training is in advance of when the skills will actually be needed.

Modular Skills Training Programme

Cummins has recently won a national award for training and it was one person's opinion that what had 'caught the eye' of the assessors was the bringing-together of several different organisations: universities, colleges, skill centres, equipment suppliers, the MSC, to provide an integrated programme with a novel approach. Valuable access was gained to up-to-date equipment of the kind which the company was likely to be using itself.

'We got the award, I think, partly because of the integration which was involved across several different activities.'

The major change-over from the Small Vee to the B-Series had been characterised by several unpredictable fluctuations, as is common in such processes:

'Products never quite run out when you've planned it. It's a very tough business decision to cut off and say - 'No more after next Tuesday.' Major customers for the B-Series might ask us to do them a small favour - 'just make us another 150 Small Vees.'

It was felt that such trade-offs were unavoidable, but that such requests did pose logistical and personal problems for the modular skills programme. Employees kept on the Small Vee production line reacted negatively to the postponement of their own training and to the sight of others receiving it.

'They thought that they might disappear into the sunset along with the Small Vee.'

The company spent much time in reassuring such employees but, as with all issues of this kind, it was remarked that the long-term plan had to be seen as,

'coming together'

before co-operation was gained. In employment terms, the situation was now better than had been predicted, and with the moving of machining processes from the old to the new building, employees were beginning to realise that the company would honour its responsibilities to continuous employment, as long as they agreed to learn new skills.

Apart from this hiccup, the modular skills training programme is progressing.

'We're well advanced into it now...'

Multi-skilling is being rapidly introduced into apprentice programmes. Areas of traditional craft skills were being developed, but there is an incorporation of whole areas of related skills. This will produce a high degree of flexibility, and a more interesting, up-to-date job from the employees' point of view. This approach was also perceived as having contributed to the training award.

The programme is being seen as 'on course',

'We were looking ahead five years when we designed this thing in 1984, and it's not too far off......but it's never static, what's said one day might be amended the next......we haven't concluded our thoughts yet.'

Work Organisation and Work Groups

'The manufacturing organisation is structured around a business concept with business managers and their teams responsible for a key function of the production process, with each business being self-sufficient in terms of operations support. This technique ensures closer involvement by all those working in each of these critical areas, providing improved control and swifter response. Teams are located on the shop-floor in offices in their relevant business area. A team concept operates within each business. Within the Darlington plant, new concepts of working practices have been evolved. The broadening of skills is encouraged with the goal of increased employee flexibility.'

This lengthy quote from the company's brochure forms a base for the following section. As described above, the new concepts of working practices and the goal of increased employee flexibility remain to be further developed in the future, but the organisation of the workforce under business managers had, it was felt, had a significant effect.

In fact later developments placed only one executive in charge of the shop. Everything which was added on, which put up the cost/unit was first questioned and if possible pruned,

'... so we cut a business out of the plan. We're a very lean organisation.'

The initial structure meant that business was concerned with the FAS from time of delivery of raw materials/sub-assemblies, until the hot test. The second took over from the hot test, to despatch. Under each business manager were two coordinators, and under the control of the coordinators were four or five teams or work groups.

When the FAS was implemented there were,

> 2 Business managers
>
> 4 Coordinators
>
> 9 Work groups/teams

Five or six years earlier, there were,

> Managers
>
> Supervisors (Superintendents)
>
> Unit coordinators
>
> Foremen
>
> Leading hands
>
> Operators - of whom there were several different categories

Thus two levels had been cut out, and all the operators had the same job specifications, instead of being allotted to such positions as operators, sub-assemblers, fork-lift drivers, line feeders, janitors, and so on. This was felt to be a much easier situation to manage. The development towards work groups was not 'an overnight thing'. It had been worked towards for three or four years, whilst the Small Vee was being assembled. Work groups, rather than functional divisions, were a company philosophy, and not just in Darlington. Work groups were seen as the cornerstone of the kind of enterprise which they were trying to bring about. What had happened at Darlington had been a coalescence of the nine planned groups around the four coordinators, so that there were in reality only five work groups, one of them having been split because it was so large. Each group, or expanded group, is responsible for some aspect of the FAS.

The introduction of the new product at Darlington is seen as the best that Cummins (as a company, not just UK) has ever done. This very success had created its own problems; volume had grown more quickly than expected; pressure points began to show up; extra capacity had to be found. All this was taking place against the background of the way in which Cummins had decided to get into certain key markets, forcing their costs down to provide a margin on the business,

'All of this was very, very demanding, doing all of these things at the same time. There was particular pressure, then, on the work teams .. we still develop the people in the team, that's a key part, but for actual day-to-day management of operations, we saw a need for a higher level of skill than we'd actually got down there. It was a dilemma, but we feel, on reflection, that we probably asked people to take on too much'.

This has removed many of the pressures from the co-ordinators. It is expected that the original plan for three business managers will be carried out in the near future, as the rise in demand continues, but until then,

'we have to come to terms with the fact that there's a supervisory/management role to be discharged here'.

'Grow Your Own Managers'

'We've always had much more success in developing our own management skills in Cummins than in bringing them in from the outside. When we've tried to do that we've failed more times than we've been successful'.

The introduction of a supervisory level has led to an examination of the kind of skills which would be needed by people assuming these responsibilities. A programme is currently being devised, where the accent is on the continuous development of technicians and technologists from the apprentice scheme (and/or with HND or first degree). This might lead to the gaining of a Master's Degree, but will in any case involve a broadening of the skills already possessed. People will be encouraged to spread out from their base discipline of mechanical engineering to include electrical engineering, and vice versa. The programme will be a demanding one. Talks were going forward with such bodies as the EITB and the MSC with a view to co-operation in developing the programme. It was hoped to establish core groups of five at each plant.

On management training generally, it would appear that the company has a good record. Once the technicians/technologists have finished their apprenticeship, many have typically followed one of two routes,

- Advanced manufacturing engineering or
- Product development.

These are both regarded as very important areas, and the record of success of various individuals, some of whom have ended up as operational managers, is impressive. These skills have been needed and available, for developing FMS control systems etc., for example, at Shotts. The company feels justly proud of these successes in, 'bringing our own people on',

but it is now perceived that these two areas have taken a disproportionate share of talented young people, and that a much larger number of,

'...our next generation of managers'

should have a broader view of Cummins, and start gaining managerial experience from actually managing down on the shop-floor, dealing with projects, and all the day-to-day problems which go along with that. It was perceived that one of the difficulties of young engineers in production areas was that the professional qualification (C Eng) was easier to get in the highly technical, specialised areas mentioned earlier.

Several companies in the North East are now thinking of ways of integrating their training resources, both at management and lower levels, and working with local companies, polytechnics and universities to develop and validate an MBA programme, as well as a Master's degree in Engineering. This development of the individual would continue as long as the person demonstrated commitment to the company.

'At the end of the day, we're growing the future management of this company. Yes we do want them to get the technical expertise from the specialist areas, but they've got to be as broadly-based as we can possibly make them.'

Team Building

There remains a training difficulty in this area which has not yet been solved,

'The problem across the whole training front has been the increasing demand for the product. We've actually got to make engines! There is a difficulty of freeing people up to go for training.'

The same person observed, however, that such programmes as NSE have played a key part in changing people's attitudes, and that the actual process of working in teams encourages people to think of the part they are playing in that team.

'This modifies behaviour in a practical sense, rather than pushing them through a four-day 'training' course.'

It was hoped, however, that courses based around the continual improvement of team working skills would take place in the future.

Learning and Generalisations from this Case

Overall the implementation of the Flexible Assembly System at Cummins' Darlington plant struck us as being a highly successful operation. Although at the time of researching the case study there were still some 'teething problems' with the FAS, and as yet it was not operating at full capacity, it was nevertheless seen as a success, both from the perspective of those implementing the system and those managing and operating it.

From our outside view it is interesting to speculate on those factors that contributed to successful implementation. The most striking contextual feature surrounding and supporting the implementation was the history of change in Cummins. Change, and the management of change, appeared to be an endemic feature of life in Cummins, and the company seems to have evolved and developed a management team with a high degree of experience and competence in this area. We were struck by the openness of managers to expose their experience to external scrutiny, and their willingness to learn from it. Managing any fairly large-scale change in this context is inevitably less traumatic than in situations where there has been little competence developed in change management. Specifically, the '30-month Sprint' to freeze costs constant promoted a major organisational reappraisal around a central set of cost, quality, delivery values in which the planning of the FAS took place. This provided a company context of 'back-to-basics thinking' which ensured the investment was closely linked to its business and manufacturing strategies and must have facilitated many of the radical changes involved in planning and implementing the FAS and the organisation to manage and operate it. In our view, the well and widely articulated values of cost, quality and delivery provided consistency, and a sense of direction, which served to lower resistance in a period of wide-ranging change.

We were also impressed by the level of integration between the design philosophies which underpinned the organisation and control of the technology deployed in the FAS, as well as organisation and control of the people employed in its operation maintenance. Much of the initial success of the Darlington plant in terms of percentage 'up-time' in comparison to its US counterpart at Rocky Mount was attributed to the control philosophy employed of simplifying the control systems as much as possible and devolving control to the lowest point possible in the system. In the AIMS system control was devolved, where possible, to the human operator, and AIMS functioned as an information system rather than a control system. This enabled rapid debugging of faults and an ability to override the system when it was judged appropriate.

Similarly, in designing the work organisation, the basic control philosophy implicit in the New Standards of Excellence initiative seems to be one of devolving control as far down the hierarchy as possible. The intention with the FAS organisation was to produce semi-autonomous work groups serviced by a coordinator. There appear to have been some initial problems in getting this sort of organisation to function. It was reported that in most cases the work groups under each coordinator had collapsed into one larger group. This

157

probably reflects the difficulty most work groups experience in changing from a culture and history where they have been heavily dependent on direct supervision, to one where they take over some of these functions for themselves. Our experience of semi-autonomous work groups in other companies suggests that their success is dependent on a relatively high initial training investment in teamworking skills, in order to incorporate those aspects of group self-management necessary for them to function without a supervisor. We understand from our discussions that Cummins did not provide this type of training support to work groups.

We were also impressed by the process of financial justification of AMT investment developed in Cummins. The attempts to quantify some of the intangible benefits of AMT reflected both a broadening of perspective in financial appraisal compared with many other companies, and also a more sophisticated grasp of business issues by engineers. However, we did feel that, despite the sophistication of the justification process, it was still firmly grounded in traditional financial appraisal techniques rather than being based on a broader multi-dimensional business analysis of how the AMT enhanced competitive position. Perhaps this type of appraisal took place in the justification process and we were not aware of it.

It was clear from our interviews that the speed of implementation of the FAS reflected not only the skilful project management of process, but also an ability to manage and coordinate change on a wide number of fronts. The project was truly conceived and managed in socio-technical terms.

- The spending of time at the front end was amply justified.
  'Things went through with far fewer hitches.'

- There must be a flexible workforce to accompany any flexible system. They were still having to work on this in Darlington.

- It must be realised that in some cases the upgrading of skills required is high,

'Things cannot just be dumped on the floor and expected to work.'

The experiences at Darlington had been useful in the development of a new system at Shotts. The right man (originally in CNC maintenance), had been recruited at the start, he had been properly trained and had passed on his expertise to others on the shop-floor, who were now,

'... running rings round the installer.'

- Spend 'as long as it takes' upon working out the control philosophy. Keep it simple, or as simple as possible.

CHAPTER 7

Rotabroach Ltd

Rotabroach is part of the Neepsend Group of Companies. It is one of eight companies in the group, and prior to the developments mentioned below, the Neepsend board had reduced the number of associated companies in the group from 48 to eight. It is a small company based in Sheffield employing about 50 people.

Rotabroach manufactures cutters and drills under a licence which is a USA patent. It is a 20 year licence and still has 15 years left to run. At the moment Rotabroach sells to 52 districts in the UK and 40 countries outside. The licence does not permit selling to China, South Korea, USA and Canada, South America or certain parts of the Pacific Basin, notably the Philippines. The rest of the world constitutes the potential market.

Because of the effects of selling under a licence, the company has a patent-protected product and consequently finds itself in a near-monopolistic position with regard to the home market. For example, Rotabroach already has 85% of the UK market for Magdrill cutters.

As a result of this, business expansion lies in exporting, and present targets are aiming for

15% growth per annum for the next three years. This will mainly be in European markets. The aim of the managing director is to dominate the market and make inroads into the existing twist drill market which the Magdrill cutter can replace, although it is more costly. The managing director works to a three year plan and, when asked why the Company had decided to invest in AMT, replied that it was for three reasons,

'Firstly, to improve the quality of the product which was a major problem for us; secondly to be the best in the world at what we do. We want to compete with the Japanese, in fact we have already seen them off in Singapore; and thirdly, thereby to ensure the future of the company.'

Overall the whole investment in new technology (see below) is seen within the company as a significant success (usual criteria, cost reduction, quality improvement, etc.). Six people have been added to the company payroll over the last two years. As well as absorbing over-manning, there has been an increase in volume during this period. Turnover has increased from £1m in 1984, to a projected £2m in March 1987. The managing director expressed the benefits of the investment in AMT in terms of both quality and quantity per person employed as 'unsurpassable' and 'more than expected' in each case. Clearly he believes that the investment has proved to be fully justified.

Rotabroach has three sales representatives organised geographically to cover the North, South and Midlands of Great Britain. The managing director covers overseas sales.

The patented product has positive and negative aspects. The patent protection results in a captured market. This market is fairly limited in that there are holes that the product can and cannot cut, and this dictates the size of the market. It is used largely in the fabrication business and in general engineering. The negative aspects are to do with quality. Old manufacturing methods use milling machines to produce the teeth of the cutter. This does not provide a sufficiently tight tolerance to ensure a high quality cutter, and results in large numbers of customer complaints. As a consequence, salesmen spent a significant proportion of their time dealing with customer complaints, rather than selling.

The Technology

In the late '70s and early '80s a decision was made (by the Management Group of Rotabroach), to invest in new technology to do several things:

- Improve quality. Some of the machines to be bought included high quality grinding machines which would allow tighter tolerances and solve the quality problem. The managing director estimated that the new machines operated at 300% better quality than the old machines. Figures are showing changes in scrap rates from 20% down to 2-5%. Accuracy is taken for granted (two-tenths of a thousandth of an inch rather than three thousandths).

- Expand capacity. The new machines were able to produce 60,000 cutters per year, as opposed to 40,000 cutters on an old machine.

- Hence expand market share. See a way forward for Rotabroach, contributing to the whole Neepsend group.

All of these have contributed to a stated desire to attack the market and maximise profits, thus exploiting the near-monopolistic situation of the company.

The detail of the implementation in technical terms has taken two and a half years from 1983. This was initially a three year plan, which set out to improve the situation regarding the cutter manufacture, then address the situation regarding production of the drill (this still has to be done) and thirdly, introduce on-line management information systems to support the marketing effort (this is being considered for 1986/87). Between 1984 and 1987 the following CNC machine tools have been installed or are ordered,

- 2 Takasawi lathes
- 3 Truflute 4-axis grinding machines
- 2 Walter 6-axis grinding machines
- 1 Bridgeport BPH 90

Two new pieces of inspection equipment have also been bought since August 1986, which give a degree of precision in checking the quality of the cutters which was unobtainable with the old machines. The effect of all this has been a vast quality improvement resulting in fewer customer complaints and a freeing up of the sales force to indulge in selling activities. This improvement in quality is the competitive edge that the company planned to obtain from the technology.

The machines were justified on the basis of increasing capacity by a half and reducing the number of operators required from seven per shift on the old machines, to two per shift on the new machines. These changes have resulted in over-manning which has been absorbed into other jobs in Rotabroach, notably inspection, grinding, warehousing (more exports) and operation of the lathes to manufacture parts which were previously bought out. Two further employees will shortly be due for retirement and now work part-time. This is also to help the over-manning situation. The total investment was £1 million over two years.

Installation and Implementation

The technical director masterminded the installation of technology. The managing director made the case to the Neepsend board and obtained their support. Having reduced from 48 companies in the group to eight, it seems that the Neepsend board were ready to indulge in such an aggressive strategy as has been adopted at Rotabroach. There is no doubt that the technical director is identified as, and identifies himself as, 'technology champion'.

The introduction of the technology has required a considerable retraining programme which has been managed by the training officer, who produced a training plan. Training has been done by the CNC machine tool manufacturing companies off-site. This has included training for production engineers, the works manager, the maintenance fitters and the operatives. Most of the training is concerned with set-up, programming and, in a more specialised way, with maintenance. However, once the operatives are familiar with tooling, and the documentation has been drawn up by the development engineer, operatives are able to use a master tape thereafter, given the limited range of parts in production at Rotabroach.

Rotabroach has adopted a 'one machine at a time' implementation strategy. This has had the benefit of allowing the impact of any one change to be absorbed, and allowing the management to learn about the problems of implementation. The experience gained in introducing the first CNC lathe, for example, is seen as invaluable by the management when introducing the second CNC lathe.

Much more flexibility of the workforce has been required than is usual, in order that they could generate their own computer programmes on the CNCs, and modify programmes as they went along. Overall this has required a higher grade of operator to be recruited. The difficulty of recruiting these people is seen by management as the most significant problem that Rotabroach has faced over the whole change. The major union is the AEU. There have been no problems here.

Payment systems work on a basic pay plus group bonus. This group bonus is paid based on outputs and the numbers of operators involved. Flexibility of the workforce and the lack of a history of piecework are seen as vital ingredients for the success of the CNC applications at Rotabroach. To emphasise this, the personnel section is gradually ensuring that everyone is trained in every job to ensure ultimate flexibility of operation. Given the limited range of parts produced, wage incentives do not have to be extremely high to attract exceptional workers. It is a matter of attracting appropriate people which keeps wage incentives quite low. Incentives are only paid for the operation of more than one machine.

The shop is divided into basic engineering units, each with a chargehand (grinding, turning, milling). Prior to the change, the shop was organised in this way, but the structure was not quite as tight (there are clear responsibilities for each of the chargehands, who receive a small supplement for undertaking these supervisory duties). A new chief inspector has been recruited for quality assurance.

The real change has been in flexibility of working, and this has been aided as with other firms by there being a climate for change within the Neepsend group, and Rotabroach in particular, within Sheffield and also nationally. A major advantage which Rotabroach had was that there was no piecework on the old system. Thus there has always been a feeling of corporateness in the manufacturing process. However, no organisational change programmes have been needed other than constant reiteration by the managing director of the value of organising to maximally exploit the technology. This important aspect of the change process is described by the managing director as an important part of his job. In such a small organisation, he is able to identify and communicate with all employees fairly easily and quickly, but emphasises the importance of constantly encouraging appropriate behaviours for managers in larger companies too, particularly using more influential and verbose employees to pass the message, not just the trades union structure. In managing the workforce currently, the managing director sees his role as instilling the need to maximise the investment in new technology in all operatives. This he does in chance encounters, formal meetings, and any other points of contact. This is probably the most important central value in driving a successful implementation in Rotabroach.

Rotabroach has always had three sales representatives. However, two new replacement representatives have been recruited, and together with the sales director and the managing director comprise the total current sales force.

Essentially the company sells through distributors and the key sales control lies in the choice of appropriate distributor. As it aims to increase its turnover by export, Europe has become its main target. The next step in technology implementation is the need to support this selling effort, and an on-line Management Information System is required in order to continually update all three directors (technical director, managing director and sales director), plus the accountant and works manager, with the current stock and manufacturing situation.

Rotabroach is essentially a service industry. It produces consumables for machine tools and as a result of this provides the best service when it can produce ex-stock, thus decreasing delivery lead-times. A revised Management Information System would allow any of the senior managers to respond immediately to customer requests in a reliable fashion, thus providing a competitive edge for Rotabroach. This may swing the lucrative twist drill market in their direction, and would comprise a final investment in new technology, enabling salesmen to attack the market even more vigorously. This has been the major effect in selling, whereby salesmen can aggressively sell rather than 'sweep-up' quality complaints.

The Future

The next major change is the Management Information System, which would cover the stock situation, costs, etc., although it is not conceived as a full-blown CIM system. Inevitably it is possible to conceive of a near-CIM situation incorporating not just CAM, but also on-line management information as a result of this. The new Management Information System under consideration would cost approximately £100,000. This may be difficult to justify to the Neepsend group board because of the intangible benefits that it produces.

The decision-making process for the MIS has been to consider in the following order:

- What is the business requirement?

- Which software would deliver the relevant information?

- On what hardware would the software run?

- What level of service and support would be offered by a supplier (there has to be someone in Sheffield - an hour away is too far)?

- The facility for change, growth and integration with other systems.

Learning and Generalisations from this Case

Rotabroach is a good example of a focused factory. It has a small range of products, low manning, and a dedicated manufacturing policy. Its markets are limited and there is no reason why it should not move towards a CIM situation in the near future. In business terms Rotabroach is in good shape to compete. Quality improvement as a business and manufacturing strategy has given the edge required. The new technology investment is fairly simple and the next step is to move toward a more integrated manufacturing and information system.

Organisationally, Rotabroach has not changed a great deal. There is some increase in flexibility on the shop-floor, and the usual associated problems of recruitment and training.

Managerially, chargehands now have a tighter role than they had before, but senior management have remained relatively undisturbed. This poses the question as to whether senior management in all small companies will be relatively undisturbed by the introduction of new technology requiring managerial integration. The lack of job differentiation, the lack of specialisation, the lack of bureaucracy, the highly organic nature of the culture in many small organisations, and the relative ease with which coordination and control are usually managed, may well be a feature of small organisations, enabling the introduction of new technology to proceed with relative ease.

In terms of implementation, the linear/learning approach adopted by Rotabroach has clearly been a success. Clearly the management had an overall vision driving the technological change which has remained relatively constant throughout the whole process, and was allied to a fixed timescale. On the other hand, rather than opting for a total and sudden change, the company has, over a relatively short time, learned how to implement successfully by introducing one machine at a time. In enabling management to learn of the technical, organisational and attitudinal issues this has been an effective strategy, and has the added advantage of spreading the cost.

References

BESSANT J & HAYWOOD B (1985) *The Introduction of Flexible Manufacturing Systems as an Example of Computer Integrated Manufacturing,* Brighton Polytechnic, Brighton

BLAKE R R and MOUTON J S (1972) *The Managerial Grid,* Gul, Houston, Texas

BODDY D and BUCHANAN D (1984) *Organisations in the Computer Age,* Gower, Aldershot

CHILDE S J (1989) 'Flexibility through the development of manufacturing infrastructures' *Proceedings of International Conference on Manufacturing Technology: its integration and management,* Sunderland, March

DEMPSEY P (1983) 'New Corporate Perspectives in FMS' in K Rathmill, Ed., *Proceedings of 2nd International Conference on Flexible Manufacturing Systems,* London, IFS (Publications) Ltd

DOSI G (1982) 'Technological paradigms and Technological Trajectories', *Research Policy*, 11,3, pp 147-162

ETTLIE J E (1988) *Taking Charge of Manufacturing*, Jossey Bass, Oxford

FLECK J (1987) 'Innovation or diffusion?' Working paper WP37/1, University of Edinburgh, Dept. of Business Studies

HAYES R H & WHEELWRIGHT S C (1984) *Restoring our Competitive Edge: Competing Through Manufacturing*, John Wiley & Sons, New York

HILL T (1985) *Manufacturing Strategy: The Strategic Management of the Manufacturing Function*, MacMillan, London

INGERSOLL ENGINEERS (1985) 'Hitech can flop in factories' *Guardian* Report of a poll commissioned by Ingersoll Engineers (12.5.85)

KUHN T (1962) *The Structure of Scientific Revolutions*, University of Chicago Press Chicago

LAWRENCE P and LORSCH J (1967) *Organisation and Environment,* Harvard University Press, Cambridge, Mass.

LIU M et al (1990) 'Organisation design for Technological Change', *Human Relations* 43,1 pp 7-22

McCRACKEN J K (1986) 'Exploitation of FMS Technology to Achieve Strategic Objectives', *Proceedings of 5th International Conference on Flexible Manufacturing Systems*, Stratford-upon-Avon, IFS (Publications) Ltd

MINTZBERG H (1979) *The Structure of Organisations,* Prentice Hall, Englewood Cliffs, N.J.

References

MINTZBERG H (1983) *Power in and around Organisations* Prentice Hall, Englewood Cliffs, N.J.

MINTZBERG H (1989) *Mintzberg on Management: inside our strange world of organisations,* Free Press, New York

MORGAN G (1986) *Images of Organisation,* Sage, London

PARNABY J (1988) 'A Systems Approach to the Implementation of JIT Methodologies in Lucas Industries', *International Journal of Production Research*, 26,3

PEREZ C (1984) 'Microelectronics, long waves and World structural change', *World Development* 13,3 pp 441-463

PEREZ C and FREEMAN C (1988) 'Structural crises of adjustment, business cycles and investment behaviour' in G Dosi et al (Eds.) *Technical Change and Economic Theory,* Francis Pinter , London

PETERS T and WATERMAN R (1982) *In Search of Excellence,* Harper and Row, New York

SCHEIN E G (1984) 'Coming to a New Awareness of Organisational Culture', *Sloan Management Review*, Winter

SCHONBERGER R J (1986) *World Class Manufacturing: The Lessons of Simplicity Applied,* Free Press, New York

SKINNER W (1985) *Manufacturing: The Formidable Competitive Weapon,* John Wiley & Sons, New York

SMITH K K (1982) 'Philosophical Problems in Thinking about Organisational Change in Goodman & Assoc.', *Change in Organisations*, Jossey Bass, Oxford

SMITH J S & TRANFIELD D R (1987) 'The Implementation and Exploitation of AMT: an Outline Methodology', Research Paper No.2, Change Management Research Unit, Sheffield City Polytechnic, later given at the Inaugural Conference of the British Academy of Management under the title: 'A Strategic Methodology for Implementing Technical Change in Manufacturing', University of Warwick

SMITH J S & TRANFIELD D R (1988) 'A Catalytic Implementation Methodology for CIM', *International Journal of CIM*, 1, 2

SMITH J S, TRANFIELD D R, BESSANT J, LEVY P, and LEY C (1990) 'Changing Organisation Design and Practices for Computer Integrated Technologies, in C Voss (Ed.) *Proceedings of the Operations Management Association conference*, University of Warwick

SORGE A et al (1982) *Microelectronics and Manpower in Manufacturing*, Gower, Aldershot

THOMPSON J (1967) *Organisations in Action*, McGraw Hill, New York

TRANFIELD D R & SMITH J S (1986) 'A Culture Change Approach to Managing Technological Innovation', Research Paper No.1, Change Management Research Unit, Sheffield City Polytechnic, later given at the 3rd International Conference on Human Factors in Manufacturing under the title 'Managing Technological Change: Tackling Taken for Granted Assumptions', IFS (Conferences) Ltd

TRANFIELD D R & SMITH J S (1988) 'Managing Rapid Change', *Management Decision Journal*, 1, 3

TRANFIELD D R, SMITH J S, BESSANT J, LEVY P, and LEY C (1990) 'Management and organisation for Computer Integrated Technologies', *International Journal of Human Factors in Manufacturing*, 1,1

TRANFIELD D R, SMITH J S, BESSANT J, LEVY P, & LEY C (1990) 'Emerging Organisation Design for Computer Integrated Technologies', *Proceedings of the International Conference on Information Technology and People*, Institution of Electrical Engineers

References

VOSS C (1985) 'Success and Failure in Advanced Manufacturing Technology' Working Paper, University of Warwick

WATERLOW G & MONNIOT J P (1986) *A Study of the State of the Art in Computer Aided Production Management in UK Industry*, ACME Directorate, Science & Engineering Research Council, Swindon

ZUBOV S (1985) 'Automate/Informate: the two faces of intelligent technology', Organisational Dynamics, Autumn